Aims and Scope: Plant and Vegetation is a Springer book series that shares current knowledge and new perspectives on global vegetation. It examines vegetation patterns and changes at all spatial and time scales – from individual plants to plant communities, and from landscapes to biomes. This series provides a source of diverse and validated information covering key issues including land cover, climate interactions, ecosystem functions, invasive species, and dynamic responses to anthropogenic influences, amongst others. It is a valuable resource for graduate students and researchers with an interest in vegetation science, geography, biodiversity, applied ecology and related disciplines. This contemporary knowledge also aims to inform managers and decision-makers involved in landscape planning, biodiversity conservation and restoration projects around the world.

Submit Your Book Proposal: Prospective authors and editors are invited to download and complete the Book Proposal Form and submit it directly to the Publishing Editor. For inquiries, please feel free to contact the Publishing Editor (contact details on the right).

Steven L. Stephenson • Rose-Marie Muzika

Forests of the Southern Appalachian Highlands and Interior Highlands of the United States

Springer

Steven L. Stephenson
Department of Biological Sciences
University of Arkansas (retired)
Fayetteville, AR, USA

Rose-Marie Muzika
School of Natural Resources
University of Missouri (retired)
Columbia, MO, USA

Some sections of chapters 7, 8, 9, and 10 are adapted from A Natural History of the Central Appalachians by Steve L. Stephenson. Copyright © 2013 West Virginia University Press.

ISSN 1875-1318 ISSN 1875-1326 (electronic)
Plant and Vegetation
ISBN 978-3-032-01154-1 ISBN 978-3-032-01155-8 (eBook)
https://doi.org/10.1007/978-3-032-01155-8

This Springer imprint is published by the registered company Springer Nature Switzerland AG
The registered company address is: Gewerbestrasse 11, 6330 Cham, Switzerland

Preface

The Appalachian Highlands and the Interior Highlands are the two largest mountainous regions between the eastern seaboard and the Rocky Mountains. The Appalachian Highlands (also referred to as the Appalachian Mountains) stretch from Alabama and Georgia to Maine, although only the portion south of approximately 39 °N latitude is considered in this book. The Interior Highlands are often referred to as the Ozark Mountains, but the latter represent only a portion of the entire region, all of which is south of 39 °N latitude. Both the Southern Appalachian Highlands (for convenience, the term "Appalachians" is used in this book) and the Interior Highlands (designated as the "Ozarks" in this book) occupy ancient landscapes that were largely unaffected by continental glaciation during the late Pleistocene. Over the past 12,000 years, the forests and other vegetation types of the two regions have undergone major changes. By about 3000 years ago, forests of the two regions had essentially reached what was their modern (at least pre-European colonization) expression. Since then, they have been subjected to increasingly greater levels of human disturbance. Disturbance was fairly minimal until the arrival of European settlers in the 1700s. Since then, clearing of land for agriculture, industry, and other types of development has impacted forests throughout both regions. In addition, logging, mining, and the introduction of the chestnut blight fungus to North America also seriously affected forests in some portions of the Appalachians and Ozarks.

The primary purpose of this book is to describe the forests and other vegetation types with a more limited distribution found today in the Appalachians and Ozarks. Forests are more than just trees, and the assemblages of other terrestrial organisms found in the two regions also are discussed. These include the plants that make up the strata of vegetation represented by small trees, shrubs/vines, herbaceous plants, and the largely ground-dwelling non-vascular plants (lichens, bryophytes, and fungi). Some of the more interesting or representative examples of each of these groups are discussed in detail. Many of the same species of plants and animals occur in both regions, but there are others that do not. The similarities and differences that exist in the two regions are discussed, and some of the now-extinct animals are described. Other topics considered in this book are the history of each region, past and present threats to the forests, the degree of protection afforded to forests, the potential effects of climate change, and places where relatively intact areas of forest still exist.

The two coauthors had the opportunity to study the forests in both regions during their careers. Stephenson did this while on the faculty of Fairmont State College (now Fairmont State University) in West Virginia and the University of Arkansas. Muzika did the same while working for the USDA Forest Service in West Virginia and as a faculty member at the University of Missouri. This book reflects their knowledge and experience. It is our hope that the information contained within will allow the reader to gain a deeper understanding of the two regions being considered.

Fayetteville, AR, USA — Steven L. Stephenson
Columbia, MO, USA — Rose-Marie Muzika

Acknowledgements

We offer our sincere appreciation to the many people who made this book possible. We especially thank those who contributed images used in this book. These included Gary Fleming, Kevin Adams, James Mitchell, Tony Collins, Brady O'Brien, Zach Bradford, Vivian Stockman, Emily Johnson, Todd Elliott, Elizabeth Koziol, Charles Garratt, David Taylor, Fred Paillet, Kim Fleming, Jason Hollinger, Irvine Wilson, Karen Willard, Tyler Reber, Chris Helzer, Michael Slay, Matt Orsie, Eva Paulhus, Randy Darrah, William Martin, Eric Isley, Susan Farrington, Denise Vaughn, and Paul Sattler. We also express thanks to a number of people who reviewed portions of the book or provided input on particular subjects. Among these were Charles Garratt, Doug Ogle, Barbara Stephenson, Gary Fleming, Mark Ford, Bill Buck, Anant Deshwal, and Tom Pauley.

The fieldwork we carried out in the Appalachians and Ozarks during our careers would not have been possible without the support of Fairmont State University, the University of Arkansas, the University of Missouri, and the United States Forest Service. The first coauthor also would like to acknowledge Dr. Harold S. Adams, who was an active collaborator in much of the fieldwork carried out in the Appalachians over a period of more than 30 years. The second author wishes to express her deepest gratitude to several colleagues who have taught and shared many details about the Ozarks over the years, including the late Dr. Richard Guyette, Denise Vaughn, Dan Drees, Susan Farrington, and countless others.

Steven L. Stephenson
Rose-Marie Muzika

Competing Interests The authors have no competing interests to declare that are relevant to the content of this manuscript.

Contents

About the Authors

Steven L. Stephenson recently retired from the University of Arkansas, where he was a Research Professor in the Department of Biological Sciences. He received his Ph.D. from Virginia Polytechnic Institute and State University and taught at Fairmont State University for 27 years before moving to the University of Arkansas. During his academic career, he has been a Senior Fulbright Scholar at Himachal University in India, a Visiting Scientist at the Australian Antarctic Division, and the William Evans Visiting Fellow at the University of Otago in New Zealand. He also received Fulbright Specialist awards to spend time at universities in India and Vietnam. In addition to his studies in the Southern Appalachians and Ozarks, Stephenson has carried out research in the forests of India, Australia, New Zealand, Thailand, Puerto Rico, and Costa Rica. He is the author or coauthor of 20 books on slime molds, fungi, and various aspects of ecology.

Rose-Marie Muzika received her Ph.D. from Michigan State University in Forestry with an emphasis in Ecology and Evolutionary Biology. For 20 years, Muzika was a Professor of Forest Ecology and Forest Health at the University of Missouri, where she also served as Department Head of Forestry and Associate Director of the School of Natural Resources. Before joining the faculty of the University of Missouri, she held positions as a research ecologist and research entomologist with the USDA Forest Service Research in Oregon and West Virginia. Since her retirement, Muzika served as director of science at the Carnegie Museum of Natural History and a senior scientist at the Powdermill Nature Reserve Field Station near Rector, Pennsylvania. During her career, she received a Fulbright Fellowship at the University of Iceland and a Bullard Fellowship in Forest Research at the Harvard Forest.

Chapter 1
The Southern Appalachian Highlands and Interior Highlands of the United States

Abstract The Southern Appalachian Highlands (referred to herein as the Appalachians) and Interior Highlands (referred herein as the Ozarks) are the two largest mountainous regions between the Atlantic seaboard and the Rocky Mountains. The forests and other types of vegetation present in these two regions occur over a landscape that varies considerably with respect to elevation and topography, and they truly can be considered "bookends of the Eastern Deciduous Forest Biome" (EDF). Braun comprehensively characterized the EDF as sufficiently similar to be considered as one vegetation type, but the latter encompasses a wide range of differences in geology, soil, vegetation, and history. As such, numerous distinctions can be made within the EDF. These are most notable in the two "bookends." These distinctions are reflected most prominently in the structure and composition of the vegetation present at a particular site. Much of the area in both regions consists of rugged dissected plateaus, but the Appalachians also contain numerous mountains with summits and ridges exceeding 1200 m. Mount Mitchell (2027 m) in North Carolina is the highest peak in all of eastern North America. In contrast, maximum elevation in the Ozarks is 839 m (Mount Magazine in Arkansas). Rocks exposed at the surface in both regions are mostly sedimentary, primarily sandstones and limestones/dolomites formed during the Paleozoic Era. The Appalachians and Ozarks were largely unaffected by "recent" continental glaciation, but disturbance factors related mostly to human activities have been and continue to be important in the two regions.

Keywords Appalachian Highlands · Appalachian Plateau · Blue Ridge · Forest types · Geology · Interior Highlands · Physiographic provinces · Ridge and Valley

S. L. Stephenson, R.-M. Muzika, *Forests of the Southern Appalachian Highlands and Interior Highlands of the United States*, Plant and Vegetation 20,
https://doi.org/10.1007/978-3-032-01155-8_1

1.1 Introduction

The Appalachian Highlands and the Interior Highlands are the two largest mountainous regions between the eastern seaboard and the Rocky Mountains (Fig. 1.1).

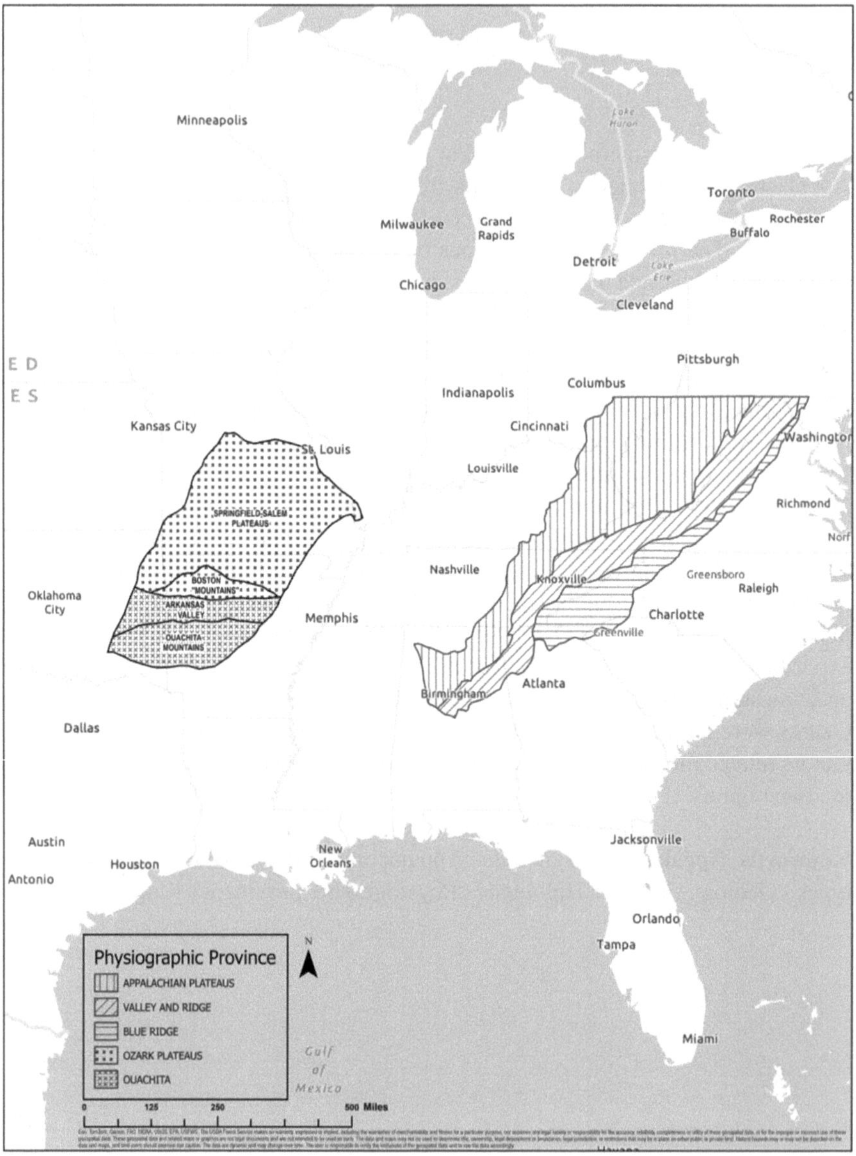

Fig. 1.1 Map of the eastern United States that shows the two regions (Appalachians and Ozarks) considered in this book. (Figure provided by Joseph Saxfield)

The Appalachian Highlands (also referred to as the Appalachian Mountains) stretch from Alabama and Georgia to Maine, but only the portion south of approximately 39 °N latitude will be considered in this book. This includes parts of Ohio, West Virginia, Maryland, Virginia, North Carolina, South Carolina, Tennessee, Kentucky, Alabama, and Georgia. The Interior Highlands are often referred to as the Ozark Mountains, but the latter represent only a portion of the entire region, all of which is south of 39 °N latitude and south. The Interior Highlands encompass portions of northern and western Arkansas, southern Missouri, eastern Oklahoma, and a small portion of extreme southeastern Kansas.

Both the Southern Appalachian Highlands (for convenience, the term "Appalachians" will be used from this point forward) and the Interior Highlands (hereafter designated by the term "Ozarks") occupy ancient landscapes that were largely unaffected by continental glaciation during the late Pleistocene. However, the glaciers did reach a small portion of northern Kentucky, and there is some evidence of alpine glaciation in the Southern Appalachians.

1.2 The Southern Appalachian Highlands (Appalachians)

The portion of Appalachian Highlands considered in this book encompasses three different physiographic provinces. These are the Blue Ridge province, the Ridge and Valley province, and the Appalachians Plateau province.

1.2.1 The Blue Ridge

The Blue Ridge physiographic province (with a total area of about 45,627 km^2), located to the east of the Ridge and Valley, consists of two rather distinct sections separated by the Roanoke River in southwestern Virginia, the southernmost stream cutting through this ridge system (Fig. 1.2). The northern section, separated from the ridges of the Ridge and Valley by the broad, flat Shenandoah Valley in northern Virginia, is a narrow (rarely reaching 22 km and often no more than 10 km) range of relatively rugged, broad-topped mountains that only occasionally exceed 1200 m. The southern section, which extends southward from the Roanoke River through western North Carolina and eastern Tennessee into southwestern South Carolina and northern Georgia, is much more extensive (with a maximum width of more than 110 km near Asheville, North Carolina) and consists of two main ranges of mountains. These are the Blue Ridge (used in the limited sense) along the southeastern border of the province and the Unaka Mountains, of which the Great Smoky Mountains are part, located to the northwest (Fig. 1.3). The region between the Blue Ridge and Unaka Mountains contains a number of elevated, relatively flat-floored basins which are separated by more or less transverse ranges of lesser mountains (e.g., the Black Mountains and the Balsam Mountains). The southern section of the

Fig. 1.2 Mountains of the northern Blue Ridge in south central Virginia. (Photo by Steven L. Stephenson)

Fig. 1.3 A portion of the southern Blue Ridge in the Great Smoky Mountains National Park. (Photo by Kevin Adams)

Blue Ridge contains the highest mountains in eastern North America, with numerous peaks exceeding 1500 m and reaching a maximum elevation of 2038 m on Mount Mitchell in western North Carolina. Most of the mountains have become rather eroded over a long period of time, with rounded summits and ridges (Fenneman 1938).

The soils of the northern Blue Ridge in Virginia are derived from a combination of crystalline igneous rocks, formed from molten magma that cooled underground, and metamorphosed rocks formed from the alteration of other rocks by heat and/or pressure. Crystalline igneous rocks are resistant to weathering. This fact, coupled with the often steep slopes, has limited soil development and depth on most sites in this portion of the Appalachians (Fleming 2012).

Fig. 1.4 View of the longitudinal ridges of the Ridge and Valley. (Photo by Charles Garratt)

1.2.2 The Ridge and Valley

The elongated, often relatively level-crested ridges of the Ridge and Valley physiographic province, which generally run in a northeast-southwest direction, form a narrow belt that extends along the western boundary of Virginia and the eastern boundary of West Virginia southward through eastern Tennessee and into northwestern Georgia and north-central Alabama (Fig. 1.4). This belt reaches a maximum width of about 105 km in northern Virginia and narrows southward.

Many of the ridges in the southern half of Virginia reach elevations in excess of 1200 m and a few exceed 1375 m. The elevation of the valley floor reaches a maximum elevation of more than 670 m near Wytheville in southwestern Virginia. The elevations of both the ridge and valley areas decline steadily from Virginia southward (Fenneman 1938; Stephenson et al. 1993). Rocks exposed at the surface on the ridges of the Ridge and Valley are all sedimentary, with sandstone, siltstone, and limestone the most prominent. Ridges are invariably capped with highly resistant sandstone, whereas less resistant limestone has been eroded away, producing the intervening valleys. This physiographic province has a total area of approximately 78,907 km^2.

1.2.3 The Appalachian Plateaus

Two physiographic regions are recognized within the Appalachians Plateaus physiographic province. The first of these is the Alleghany Plateau (Fig. 1.5) in western West Virginia and eastern Ohio. The second physiographic region is the Cumberland Plateau (Fig. 1.6) located to the south, with most of it occurring in eastern Tennessee and western North Carolina. Collectively, these two regions have a total area of about 134,962 km^2, which is larger than the combined total for the Blue Ridge and

Fig. 1.5 A portion of the Alleghany Plateau in western central West Virginia. (Photo by Steven L. Stephenson)

Fig. 1.6 A portion of the Cumberland Plateau in eastern Kentucky. (Photo by Eric Dodd)

Ridge and Valley physiographic provinces. The Appalachian Plateaus province consists of series of rugged dissected plateaus located west of the Ridge and Valley Province. The eastern edge of the province is the highest part and has the general appearance of a mountain range due to the presence of an abrupt and steep escarpment known as the Allegheny Structural Front or more commonly, the Allegheny Front. These more mountainous eastern portions of the Alleghany Plateau are often referred to as the Alleghany Mountains. In West Virginia elevations reach 1482 m at Spruce Knob, but several other summits reach 1400 m (Stephenson 1993). From West Virginia to Tennessee, elevations drop to about 900 m and continue trending downward to 300 m in northern Alabama. On the western side of the plateau, elevations are less than 300 m in the north, increase to about 600 m in Kentucky, and then drop to 150 m in northwestern Alabama.

A large portion of the Appalachian Plateaus province is underlain by relatively horizontal rock layers consisting of sandstone, limestone, shale, siltstone, and mudstone. The soils derived from the rocks are often thin, typically rocky, and nutrient-poor. Intermixed with rock layers in the region are coal seams that were formed

320 million years ago during the Carboniferous Period. Various human activities associated with the exploitation of coal have had a major ecological impact, a topic to be discussed in a later chapter.

1.3 The Interior Highlands (Ozarks)

The Ozarks occupy a much smaller area than the Appalachians, with most of it falling within southern Missouri and western Arkansas. This region consists of two physiographic provinces—the Ozark Plateaus province and the Ouachita province. Each province consists of two geologically distinct portions. The Springfield-Salem Plateaus and Boston Mountains fall within the Ozark Plateaus province and the Ouachita province includes the Arkansas Valley and the Ouachita Mountains.

1.3.1 The Springfield-Salem Plateaus

The Springfield-Salem Plateaus portion of the Ozark Plateaus province is an uplands area of low relief and elevations of 300–460 m (Fig. 1.7). Although considered one section, the Salem Plateau and Springfield Plateaus are distinct. The latter is underlain with abundant limestone and chert, and karst features are abundant. It is an area of numerous caves and springs. The Salem plateau is north and east of the Springfield Plateau and represents the largest geographic reach of the Ozarks. The St. Francois Mountains, nestled within the Salem Plateau, in the northeastern portion of this section in Missouri, consists of low hills, knobs, and basins with bedrock of Precambrian age from ancient volcanic activity. The billion-year-old rocks have been uplifted and the overlay strata eroded resulting in unique igneous rock types and geological formations. The St. Francois Mountains reach a maximum elevation of 540 m, which is only 150–240 m above surroundings regions of the plateau. This relatively small area of the Ozarks is thought to represent the only part of the midwestern

Fig. 1.7 A portion of the Ozarks in southern Missouri. (Photo by Rose-Marie Muzika)

Fig. 1.8 The Boston Mountains in northwestern Arkansas. (Photo by Steven L. Stephenson)

United States that was not submerged by the shallow seas that covered the region during the Paleozoic and Mesozoic Eras.

The Boston Mountains portion (Fig. 1.8), located south of the Springfield-Salem Plateaus, is higher and more rugged than the Springfield and Salem Plateaus. There are narrow steep-sided valleys created from stream erosion. Elevations in the Boston Mountains reach 600 m in some places. These mountains rise abruptly from the Arkansas Valley section of the Ouachita Mountains province to the south. The Boston Mts and the Springfield plateau reach into Oklahoma.

1.3.2 The Arkansas Valley

The Arkansas Valley, which occupies the area between the Boston Mountains and the Ouachita Mountains, consists of flat lowlands with some rolling hills and is interrupted by three isolated flat-topped mountains. These are Mount Magazine (839 m), Mount Nebo (539 m), and Petit Jean Mountain (744 m). The Arkansas Valley is approximately 50–65 km wide, and the Arkansas River, a major tributary of the Mississippi, flows through it. Much of the land in the valley has been converted to agriculture and thus is now of considerable economic importance.

1.3.3 The Ouachita Mountains

The Ouachita Mountains, often referred to simply as the Ouachitas, consist of a series of mostly east-west trending ridges separated by narrow to broad valleys. The layers of rock that make up the ridges are often folded and faulted and thus comparable to some areas of the Appalachian Mountains. The highest point is Rich

Mountain at 817 m. Rocks exposed at the surface in this province are mostly shales, sandstone, and chert, with limited areas of limestone.

The western boundary of the is not clearly defined since it slopes gently down to the relatively level terrain of the Ozark Plains section of the Interior Lowlands physiographic province. To the east, the Interior Highlands transition to the broad flat Mississippi Alluvial Plain and to the south to the low hills of the Athens Plateau.

1.4 General Aspects

In general, although the Ozarks are smaller with respect to total area, lower in elevation, and with less relief than the Appalachians, the landforms of the two regions are not too dissimilar (Bryant et al. 1993). Rocks exposed at the surface in both regions are mostly sedimentary (primarily sandstone and limestone/dolostone), which is not surprising since they share a common origin. They are largely sedimentary and were formed during the Paleozoic era more than 250 million years ago when the material of which they consist was deposited on an ancient sea floor. However, the landscape of the Appalachians includes more rock types (siltstones and mudstones) along with some areas of igneous and metamorphic rocks.

The occurrence of much higher elevations and the resultant cooler temperatures in the Appalachians does allow for the presence of organisms whose distributions are centered at more northern latitudes. Overall, precipitation levels are generally lower in the Ozarks, and the drier climate results in a greater areal extent of open, non-forested areas. These and other climatic aspects of the Appalachians and Ozarks will be discussed in the next section.

1.5 The Climate

Differences in elevation, topography, and latitude result in considerable variation throughout the Appalachians and Ozarks. This is particularly true for the Appalachians because of the very wide range in elevation (>2000 m on the very highest summits). In general, temperatures, precipitation, and the length of the growing season increase from north to south in both regions. In the Ozarks, there is also a trend of decreasing precipitation from east to west. Temperatures for the coldest and warmest months of the year at the northern boundary of the Ozarks average about 0.5 °C (January) and 6.2 °C (July) lower than the corresponding values for a comparable elevation at the southern border of the region. Average highs in July are almost 9.0 °C higher near the southern boundary. For the Appalachians, average temperatures are about 4.5 °C (January) and 2.8 °C (July) lower for the northern boundary when compared to values the same elevation near the southern boundary. Average highs in July are about 3.9 °C higher in the south. The differences in temperature are much greater when the highest elevations of the region are considered particularly for the Appalachians, whereas the elevational gradient in the Ozarks does not exert much of an influence.

According to the classification of Thornthwaite (1948), the climate of both the Appalachians and Ozarks is humid mesothermal, with precipitation adequate for all seasons. However, the highest elevations in the Appalachians are considered as perhumid because of the high levels of precipitation (>200 cm/year) they receive, which are the highest in the United States outside of the Pacific Northwest (White et al. 1993). Precipitation levels were likely to have been even higher when the Wisconsin glacier was melting 10,000–12,000 years ago. In contrast, precipitation totals at the southwestern boundary of the Ozarks and in some portions of the Ridge and Valley province of the Appalachians are <100 cm/year. Slope aspect plays a critical role in influencing vegetation patterns in both regions, although more obviously so in the Ozarks. Slopes with a northern exposure have a microclimate that is typically cooler than that of slopes with a southern exposure in the same general locality. On steeper slopes, this difference can be enough to result in compositional differences in the vegetation present. The same phenomenon exists in mountain valleys as a result of cold drainage from surrounding slopes. In these situations, the microclimate of valley floor can be appreciably cooler than that of the slopes above it.

1.6 The Vegetation

The temperate deciduous forests that represent the dominant vegetation of both the Appalachians and Ozarks are mostly made up of broadleaf trees (Fig. 1.9), but there are some areas occupied by coniferous forests. Many of the same species of trees are shared in common, but some species are limited to one region or the other. This is especially true for the Appalachians, where the cool, moist conditions found at

Fig. 1.9 Deciduous forest in the Ridge and Valley of southwestern Virginia. (Photo by Charles Garratt)

higher elevations are favorable for trees such as spruce and fir (Adams and Stephenson 1991; Stephenson 1993). In contrast, conifers in the Ozarks tend to be limited to pine (*Pinus* spp.) and cedar (*Juniperus*), two genera that are associated with dry conditions. The same general patterns of distribution are followed by secondary strata of vegetation such as shrubs and herbaceous plants, with some species common to both regions while others are found only in the Appalachians or Ozarks. There are also notable distinctions within the same genus. For example, eastern witch hazel (*Hamamelis virginiana*) occurs in the Appalachians and Ozark witch hazel (*Hamamelis vernalis*) is found in the Ozarks.

Although forests represent the dominant vegetation, non-forested areas occur in both regions. These are more extensive in the Ozarks than the Appalachians, especially towards the western boundary as the region grades into the great plains. Non-forested areas in both regions represent a number of different types of habitats. Non-forested habitats are covered later in this book.

The most widely accepted classification of the forests of eastern North America is probably that of E. Lucy Braun (1950), who did extensive field work throughout this entire region during the 1920s and 1930s. This is important because she was able to observe the forests of some portions (especially the Cumberland Mountains) while they were still relatively undisturbed by logging and other human activities. As delimited by Braun, the Ozarks fall within the southern division of the Oak-Hickory Forest Region, the delineation corresponding to the glacial boundary. As characterized by Küchler (1964), who also described forest types in eastern North American, forests of the region are composed of medium tall to tall broadleaf deciduous trees, with white oak (*Quercus alba*), northern red oak (*Q. rubra*), black oak (*Q. velutina*), shagbark hickory (*Carya ovata*), and bitternut hickory (*C. cordiformis*) as the usual dominants. Oak-hickory forests also occur in some areas in the Appalachians, where another species of oak, chestnut oak (*Q. montana*), is also present (Skeen et al. 1993). The two species of hickory listed above also occur in the oak-hickory forests of the Appalachians, but two other species of hickory, pignut hickory (*C. glabra*) and mockernut hickory (*C. alba*), are just as likely to be important in some areas.

Braun (1950) placed a major portion of the Appalachians within her Oak-Chestnut Forest Region, but this name is no longer applicable because of the nearly complete elimination of American chestnut (*Castanea dentata*) by the chestnut blight fungus during the first half of the twentieth century. As such, it is now more appropriate to refer to the Oak-Chestnut Region as the Mixed Oak Region (Küchler 1964). Braun recognized three subdivisions of the Oak-Chestnut Region. These are the Southern Appalachians, the Northern Blue Ridge, and the Ridge and Valley. The western boundary of the Oak-Chestnut Region roughly follows the state line between Virginia and West Virginia for several hundred km and then straddles the state line between North Carolina and Tennessee. Braun considered the region west of the Oak-Chestnut Region as the Mixed Mesophytic Forest Region. The forests of this region are best developed in western North Carolina, eastern Tennessee, and eastern Kentucky. These forests are among the most biodiverse temperate forests in the world, and as many as 30 different species of trees can occur in the forest canopy. The high biodiversity of this region is also reflected in the species richness and abundance observed for other groups of organisms.

The highest elevations in the Appalachians (generally above 1350 m but lower in the north) support a very different type of forest than is found at lower elevations. This type of forest is largely dominated by evergreen, needle-leaved trees (conifers) rather than broadleaf trees (Fig. 1.10). The most important and abundant species present is red spruce (*Picea rubens*), which at some localities occurs in association with Fraser fir (*Abies fraseri*) in the southern portions of the Appalachians and balsam fir (*Abies balsamea*) in the northern portions of the Appalachians. Hemlock (*Tsuga canadensis*) also can be present. These high-elevation subalpine coniferous forests typically contain an admixture of broadleaf trees in addition to the conifers. In contrast to other forests in the Appalachians, the forests found at higher elevations are not very biodiverse, and at a given locality no more than two or three species of trees are present. This high-elevation forest has no counterpart in the Ozarks.

Vegetation communities in the Ozarks are strongly influenced by geological parent material, aspect, and elevation, despite a shorter range of elevation than the Appalachians. Although dolomites and limestones (carbonate rocks) prevail, there are smaller areas of igneous rocks and sandstones, which create distinct vegetation types. Generally, productivity is lower for sites in the Ozarks than in the Appalachians. Consequently, the diversity and composition of the vegetation differ between those two regions, and overstory species richness is generally lower in the forests of the Ozarks.

Braun considered the Ozarks as a subunit of the Southern Division of the Oak-Hickory Forest Region. The other subunit of the Southern Division is what she referred to as the Forest-Prairie transition area. The species richness of oaks is notable in the Ozarks, and the number of species presents is a distinguishing factor between the northern and southern divisions of the Oak-Hickory Forest Region. No fewer than 15 species of oak could be found in the Ozarks, including the rare maple leaf oak, *Quercus acerifolia*, found only in the Oauchita Mountains of Arkansas.

Fig. 1.10 Red spruce forest in the Great Smoky Mountains National Park. (Photo by Fred Paillet)

Species of pine generally increase southward in the Ozarks. In the Missouri Ozarks, shortleaf pine (*Pinus echinata*) is found in woodlands and savannas—areas that might have had frequent fire historically—as well as in closed canopy forests where it co-occurs with oak (Steyermark 1940). A once dominant species in the Ozarks of Missouri, extensive logging from the late nineteenth century to the early twentieth century reduced this species from a purported 26,000 km^2 to no more than 2000 km^2. After logging, oaks soon replaced much of the landscape once covered by pine (Benac and Flader 2004). This region might be considered as much oak-pine as oak-hickory.

The Boston Mountains support some species from the Appalachian flora that are absent from the Springfield-Salem Plateau. Beech (*Fagus grandiflora*) and cucumber magnolia (*Magnolia acuminata*) are two notable examples and their presence reflects the more mesic conditions that exist in the Boston Mountains. The forest vegetation of the Ouachita Mountains is more similar to that of the Ozarks in that pine becomes dominant.

References

Adams HS, Stephenson SL (1991) High-elevation coniferous forests in Virginia. Va J Sci 42:391–399

Benac D, Flader S (2004) History of Missouri forests in the era of exploitation and conservation. In: Spetich MA (ed) Upland Oak ecology symposium, United States Department of Agriculture Publication SRS-73. United States Department of Agriculture, Asheville, NC, pp 36–41

Braun EL (1950) Deciduous forests of Eastern North America. Blakiston, Philadelphia

Bryant WS, McComb WC, Fralish JS (1993) Oak-hickory forests (western mesophytic/oak-hickory forests). In: Martin WH, Boyce SC, Echternacht AC (eds) Biodiversity of the Southeastern United States: upland terrestrial communities. John Wiley and Sons, New York, pp 143–201

Fenneman NM (1938) Physiography of the eastern United States. McGraw-Hill, New York

Fleming GP (2012) The nature of the Virginia flora. In: Weakley AS, Ludwig JC, Townsend JF, Crowder B (eds) Flora of Virginia. Botanical Research Institute of Texas Press, Fort Worth, pp 24–75

Küchler AW (1964) Potential natural vegetation of the conterminous United States, American Geographical Society Special Publication No. 36. American Geographical Society, New York

Skeen JN, Doerr PD, van Lear DH (1993) Oak-hickory-pine forests. In: Martin WH, Boyce SC, Echternacht AC (eds) Biodiversity of the Southeastern United States: upland terrestrial communities. John Wiley and Sons, New York, pp 1–33

Stephenson SL (1993) Upland forests of West Virginia. McClain Printing Company, Parsons, WV

Stephenson SL, Ash AN, Stauffer DF (1993) Appalachian oak forests. In: Martin WH, Boyce SC, Echternacht AC (eds) Biodiversity of the Southeastern United States: Upland Terrestrial Communities. John Wiley and Sons, New York, pp 255–303

Steyermark JA (1940) Studies on the vegetation of Missouri. I. natural plant associations and succession in the Ozarks of Missouri. Field Mus Nat Hist Bot Ser 9:349–475

Thornthwaite CW (1948) An approach towards a rational classification of climate. Geogr Rev 38:55–94

White PS, Buckner ER, Pittillo JD, Cogbill CV (1993) High-elevation forests: northern hardwoods forests and associated communities. In: Martin WH, Boyce SC, Echternacht AC (eds) Biodiversity of the Southeastern United States: Upland Terrestrial Communities. John Wiley and Sons, New York, pp 305–337

Chapter 2
History of Appalachian and Ozark Forests

Abstract The forests of the Ozarks and Appalachians have undergone considerable change over time. The landscape 12,000 years ago, as the Pleistocene Epoch came to a close, was more similar to the forests that currently exist at high latitudes today than those currently found in the two regions. Native Americans first arrived more than 14,000 years ago, when elements of the Pleistocene megafauna were not yet extinct. Although their numbers were low, artifacts left behind by these early Native Americans recovered from at a number of sites in the Appalachians and Ozarks date back to 12,000–14,000 years ago. The impact of Native Americans on the landscape was likely very minimal until about 3000 years ago when what is known as the Woodland Period began. By that time, the forests of the two regions had essentially reached what was their modern (at least pre-European colonization) expression. The purpose of this chapter is to describe the history of the Appalachians and Ozarks over the past 20,000 years, with a particular focus on the human inhabitants of the two regions.

Keywords Civil War · Fossils · History · Holocene · Native North Americans · Pleistocene · Pollen data · Woodland Period

2.1 The Pleistocene Epoch

It is not known when Native Americans first reached either the Appalachians or the Ozarks, but it took place more than 14,000 years ago. Radiocarbon dates obtained for plant material associated with human artifacts collected from Cactus Hill, an archeological site in southeastern Virginia located only about 100 km east of the regions considered in this book, suggest that it was occupied at least that long ago (Wagner and McAvoy 2004). Artifacts recovered from the Meadowcroft Rockshelter, an archaeological site in southwestern Pennsylvania just north of the region considered in this book, have been assigned a similar or even earlier date. Tool-like bones thought to have been made by Native Americans were recorded from the Saltville

S. L. Stephenson, R.-M. Muzika, *Forests of the Southern Appalachian Highlands and Interior Highlands of the United States*, Plant and Vegetation 20,
https://doi.org/10.1007/978-3-032-01155-8_2

Valley in the Ridge and Valley of Virginia, and these were radiocarbon dated to approximately 14,500 years ago. At the Mastodon State Historic Site in southeastern Missouri, a stone spear point was found directly associated with the fossil bones of a mastodon. The latter has considered to date back to about 12,000 years ago. The Big Eddy site in the Ozarks of southwestern Missouri has yielded artifacts that date back at least 12,000 years and possibly as much as 14,000 years (Fig. 2.1).

Native American artifacts from what is known as the Dalton culture and dated to about 10,500 years ago are also known from a site in northeastern Arkansas just east of the eastern boundary of the Ozark region as delimited in this book (Fig. 2.2).

The world these first Native Americans inhabited was very different from what it is today. Twelve thousand years ago marked the end of the Pleistocene Epoch, which had begun about two and half million years earlier. The Pleistocene was a time internal in the earth's history that often is referred to as the "Ice Age" because of the occurrence of a number of cycles of continental glaciation during which much of Canada and the northern portion of the United States was periodically covered by huge sheets of ice. The last major advance of a major continental glacier, which has been named the Wisconsin glacial episode, was ending about 12,000 years ago. At its maximum extent between 20,000 and 25,000 years ago, the Wisconsin covered a total area of 13,000,000 km^2 and was more than 3000 m thick in some places. Although very little of the Appalachians and Ozarks was ever covered by ice during this time, the presence of the large mass of ice just to the north certainly caused temperatures to be much lower and precipitation to be much higher than today's climate. As a result of these lower temperatures, the vegetation of the two regions at the end of the Pleistocene would have been vastly different from what it is today.

Fig. 2.1 Native American point from the Big Eddy site in southeast Missouri. (Photo courtesy of Archaeological Research, Missouri State University)

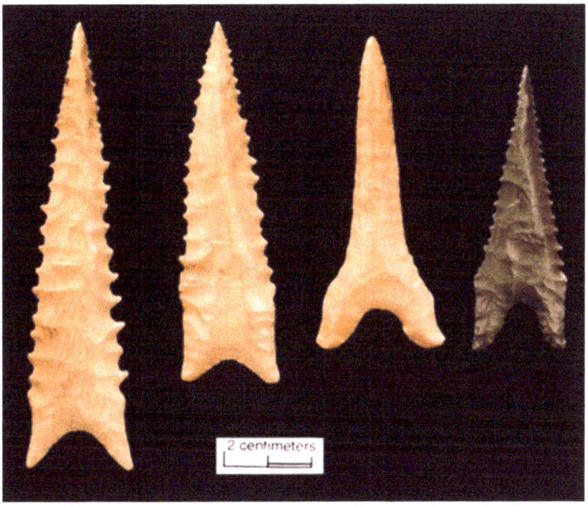

Fig. 2.2 Dalton culture points from the Sloan site in northeast Arkansas. (Photo courtesy of Arkansas Archeological Survey)

2.2 The Holocene Epoch

The end of the Pleistocene Epoch about 12,000 years ago marked the beginning of the Holocene Epoch, which extends to the present day. Reconstruction of the vegetation that existed in the Appalachians and Ozarks at different time intervals both before and during the Holocene is based largely on fossil pollen assemblages obtained from bottom sediments of bogs, swamps, and ponds. The fossil pollen assemblage from a particular locality can be radiocarbon-dated and then the plant taxa represented in the assemblage identified on the basis of morphological features of pollen grains. In many cases, the pollen grains of trees are distinctive enough to identify most taxa to the level of species. The data available for one locality represent one piece of a puzzle. It is necessary to assemble the pieces from as many localities as possible to get some idea of the picture on the completed puzzle. Unfortunately, localities where fossil pollen assemblages have been preserved are uncommon, and the intervals of time they represent only portions of the Holocene. Moreover, while localities where conditions were favorable for the preservation of pollen are scattered throughout the Appalachians, they are relatively uncommon in the Ozarks.

Delcourt and Delcourt (1981) used fossil pollen data from more than 100 localities to develop a series of maps showing the patterns of vegetation that existed in eastern North America over the past 40,000 years. A number of these localities are located within the region defined as the Appalachians in this book. As such, there is enough information available to describe in at least general terms what this region was like during a considerable length of time in the past.

The Shenandoah Valley is located between the Blue Ridge province and the easternmost ridge of the Ridge and Valley province. The valley, most of which in within the state of Virginia, is about 240 km long and approximately 40 km wide. Pollen assemblages obtained from two ponds in the southern portion of the valley provide data relating to what the vegetation was like during the late Pleistocene and early Holocene (Craig 1969). Approximately 12,700 years ago, conifer woodlands consisting of pine and spruce apparently represented the dominant forest vegetation, but the relative abundance of pollen from herbaceous plants suggests some non-forested areas of the valley also existed. By about 9500 years ago, a shift from spruce-pine to oak appears to have occurred, and this was followed by a transition to oak-pine. The pines involved were species that still occur in the region and not the species (jack pine [*Pinus banksiana*]) that cooccurred along with spruce in the late Pleistocene.

One of the few and certainly the most important fossil pollen locality in the Ozarks is Cupola Pond, a small permanent, sinkhole pond in southern Missouri (Jones et al. 2017, following Smith 1984). Analysis of pollen profiles obtained from this pond provides data on the surrounding vegetation dating back about 19,800 years. This date falls between the last glacial maximum (i.e., when the Wisconsin glaciation reached its greatest extent) and the beginning of the Holocene. The pollen data indicate that the forests which surrounded this site almost 20,000 years ago were dominated by spruce (*Picea*) and/or pine (*Pinus*), along with a small admixture of fir (*Abies*). These forests persisted, although gradually decreasing in extent, until about 11,000 years ago when spruce, fir, and pine disappeared from the pollen record. Oaks (*Quercus*) also were present in the oldest strata but only minimally; however, oaks displayed a notable increase by 14,500 years ago and became dominant by 11,500 years ago. About 1800 years ago, pine regained dominance. Undoubtedly, this was a different pine than represented in the oldest sediments, although both were identified as members of the genus *Pinus*, specifically the subgenus *Pinus*, the group known as hard pines. The latter group includes jack pine, the likely common pine thousands of years ago, as well as shortleaf pine, the current dominant pine of the Ozarks. The now widespread oak-pine forests of the Ozarks began their development 1500 years ago. Oak dominance became increasing more important throughout the latter part of the Pleistocene and this trend continued on through the Holocene. Other broadleaf trees, most notably ash (*Fraxinus*) and hophornbeam (*Ostrya*) and/or hornbeam (*Carpinus*) [the pollen of members of these two genera cannot be reliably distinguished] reached peaks about 14,000 and 12,000 years ago, respectively. Both of these taxa are now found throughout the eastern United States but do not occur far enough north to co-occur with spruce. However, this must have been the case at the Cupola Pond locality, where a transition from spruce forests to broadleaf forests appears to have been taking place.

2.3 The Early Landscape

Twenty thousand years ago, the landscape immediately south of the glacier would have been rather open, with a few small trees, scattered low-growing plants, carpets of moss, and some bare areas (Coope 1968). This type of vegetation would have been similar in appearance to the tundra-woodlands found in some areas of northern and western Alaska today (Fig. 2.3). Located to the south of these tundra woodlands were coniferous forests with white spruce (*Picea glauca*), black spruce (*P. mariana*) and/or spruce-jack pine (*P. banksiana*) present. This type of forest now occurs over much of Canada and extends as far west as Alaska. Forests where oaks (*Quercus*) were a major component occurred even farther south (Delcourt and Delcourt 1981). The extent of these vegetation types varied rather greatly with elevation, with the more northernmost types persisting longer at higher elevations, especially in the Appalachians. Moreover, each of the types gradually shifted northward or upward when possible as the glacier melted. Apparently in response to generally drier conditions, forests of the Ozarks were more open and savanna-like than those of the Appalachians during the early part of the Holocene. In the regions south of the Appalachians and Ozarks, various broadleaf species other than oak such as elm (*Ulmus*), maple (*Acer*), and hickory (*Carya*) persisted in refugial areas where growing conditions were more favorable.

The much higher elevations in the Appalachians resulted in a different response of vegetation to the presence of the glacier. At the peak of glaciation, alpine tundra occurred at the very highest elevations in the Appalachians, and geological evidence suggests that small accumulations of ice had formed at a few places well below the southern margin of the main ice sheet. Moreover, small-scale geological features such as patterned ground and block fields characteristically associated with periglacial environments have been reported from a number of high-elevation localities in the Appalachians (Clark and Ciolkosz 1988). At elevations below tundra, a subalpine zone probably included alpine meadows (White et al. 1993). Forests consisting mostly of spruce covered much of the rest of the region below about 1500 m at this time. As the glacier melted and temperatures increased, the areas of tundra disappeared and the spruce forests migrated upslope, where they still persist today. Although mostly spruce, fir now occurs on some of the higher summits in the southern portion of the Blue Ridge section of the Appalachians and at a few localities further north on the Appalachian Plateau. Interestingly, the lower elevational limit

Fig. 2.3 Spruce woodland in central Alaska. (Photo by Steven L. Stephenson)

for spruce forests today is thought to have been about the same (approximately 1500 m) as the upper limit during the glacial maximum.

2.4 The Holocene Climatic Optimum

During the period of time referred to as the Holocene climatic optimum (also referred to as the Hypsithermal Period), which extended from about 8000–5000 years ago, global temperatures are thought to have been 2.5 °C higher than they are today and also higher than they had been during the earlier part of the Holocene. This resulted in some significant changes in the pattern of vegetation that existed prior to Holocene climatic optimum. One consequence of the higher temperatures and resultant drier conditions was an eastward shift in the boundary between the western prairies and eastern broadleaf forests. Prairies probably reached their easternmost point about 7000 years ago. Forests across the region of the United States that encompassed the Appalachians and Ozarks were mostly dominated by oaks and hickories. Trees adapted to moist sites retreated to higher elevations or to topographic situations (e.g., steep-sided valleys) where environmental conditions were more favorable. Oak savannas and woodland communities occurred at the prairie-forest boundary, and open grassy areas became more frequent on the eastern side of the boundary. Pine forests expanded in areas south of the Appalachians and Ozarks, especially in the case of the former (Owen 2012).

As mentioned earlier, the spruce forests which had disappeared in the Ozarks by about 11,000 years ago (Jones et al. 2017) but still remained in the Appalachians were displaced upslope by as much as 380 m to where the cool, moist conditions required for the survival of spruce still existed. In the Ozarks, there was no elevational equivalent for spruce (or fir) to persist. On those mountains whose summits did not reach elevations high enough, this forest type would have been extirpated (Whittaker 1956). However, a drop in global temperatures after the Holocene climatic optimum ended allowed spruce to reclaim much of its former position. However, unless there was an available seed source, spruce would not have been able to reestablish itself on summits where the species had been eliminated. Samples of subfossil oak wood from northern Missouri have been used to reconstruct climate. This wood, which included oak, spruce, and fir, with some samples dating from approximately 14,000 years ago, revealed that climate change in mid continental North America was synchronous with the North Atlantic, and that climate variation had a distinct effect on tree growth and presumably species success and colonization (Guyette et al. 2006).

During the 5000 years since the end of the Holocene climatic optimum, the pattern of vegetation across the Appalachians and Ozarks gradually developed into what was present when Europeans first arrived in the two regions. Over that period, pine became more important in the southern portion of the Ozarks, oak-chestnut (*Castanea dentata*) forests became dominant over portions of the Appalachians (especially the mountain slopes of the Ridge and Valley), and species-rich mixed forests expanded out from the Cumberland Plateau (Hinkle et al. 1993). Pine and oak "settled in" to become dominant components of Ozark forests.

2.5 The Pleistocene Fauna

Twelve thousands years ago, the fauna of Appalachians and the Ozarks included a number of large mammals that would soon go extinct (Grayson 1991). These large mammals, referred to as the Pleistocene megafauna, included such huge examples as the mammoth (*Mammuthus columbi*), mastodon (*Mammot americanum*) and giant ground sloth (*Megalonix jeffersonii*). The latter was the first fossil animal described (in 1799) from the United States. The description was based on fossils found in a cave in West Virginia.

Fossils discovered at what is now Big Bone Lick State Park in northern Kentucky and near the town of Saltville in southwestern Virginia provide a glimpse into the animal life in the Appalachians during the late Pleistocene. In addition to the examples mentioned above, these two sites yielded fossils of other animals such as the sabre-tooth cat (*Smilodon fatalis*), caribou (*Rangifer tarandus*), giant beaver (*Castoroides ohioensis*), musk ox (*Ovibos moschatus*), ancient bison (*Bison antiquus*), and a species of horse (*Equus scotti*). These animals had been attracted to the site because of the presence of a naturally occurring salt lick, and some of them were trapped there by the soft marshy ground surrounding the salt lick. A full-size mastodon skeleton is on display at the Museum of the Middle Appalachians in Saltville, Virginia (Fig. 2.4).

Late-Pleistocene fossils recovered from caves and other miscellaneous sites in the northern Ozarks in Missouri (King 1973; Hawksley 1986) include, in addition to the mammals listed above, the dire wolf (*Aenocyon dirus*), short-faced bear (*Arctodus simus*), a llama-like camel (*Camelops*), and the giant armadillo-like glyptodont (*Glyptodon*). By the end of the Pleistocene, all of these animals had gone extinct except for the musk ox and caribou, which now occur no closer than Alaska

Fig. 2.4 Mastodon skeleton on display at the Museum of the Middle Appalachians in Saltville, Virginia. (Photo by Steven L. Stephenson)

and Canada in North America. In addition to these larger mammals, there were numerous species of small mammals (referred to as the microfauna) present in the Appalachians and Ozarks during the same time period. These were clearly dominated by various forms, in particular the voles and lemmings, now characteristic of the boreal coniferous forests found hundreds of kilometers to the north. For example, the yellow-cheeked vole (*Microtus xanthognathus*), which was exceedingly common throughout the Appalachians and Ozarks in the late Pleistocene, is found today no closer than the west coast of Hudson Bay in northern Canada (Guilday et al. 1977; Guilday 1982).

The mass extinction of the megafauna has been attributed to several factors, including overhunting by early Native Americans, but climate change associated with the retreat of the glaciers was certainly a major factor. Nevertheless, these early human inhabitants of the Appalachians and Ozarks undoubtedly hunted many of the animals present in the two regions. However, killing one of the really large and dangerous animals was likely an infrequent occurrence.

2.6 Late Holocene to the Present Day

By the middle of the Holocene, the forests of the Appalachians and Ozarks had largely become what the Europeans encountered when they first reached the two regions in the 1600s. By the mid-Holocene, all of the large mammals of the Pleistocene had been gone for thousands of years. Only a few of the mammals still found in the Appalachians and Ozarks were heavier than 20 kg. These were white-tailed deer (*Odocoileus virginianus*), elk (*Cervus elaphus*), mountain lion (*Felis concolor*), timber wolf (*Canis lupus*), black bear (*Ursus americanus*), and the American bison (*Bison bison*). The latter, which was represented by a smaller variant (the "wood buffalo") in the Appalachians, was more common in the Ozarks than the Appalachians. This was probably due in part to the fact that the Ozarks never had the almost complete forest cover that is characteristic of the Appalachians, and the western edge of the Ozark forests intergrades with the grasslands of the Great Plains. Except for various types of special habitats, the forests of the Appalachians and Ozarks were ecologically rather homogeneous over large areas of the landscape.

The early Native Americans lived as hunter-gatherers who left little evidence of their presence. Living in small groups, they moved from place to place, going wherever local conditions provided what was needed in order to survive. Before the extinction of the megafauna, artifacts and fossils recovered at some sites suggest that they sometimes hunted large animals such as the mammoth and mastodon. However, for the most part they hunted many of the same smaller animals that still survive today. The most important mineral resource exploited by early Native Americans was chert or flint, which was used for fashioning stone tools. These cryptocrystalline silica dioxide-based materials are found in both the Appalachians and Ozarks. Other non-chert materials used for making stone tools include sandstone, quartz, quartzite, and siltsone (Lesser 1993). Some stone tools found in the

Fig. 2.5 Rock shelter located on the Ozark Plateau in south central Missouri. (Photo courtesy of Cultural Resource Analysis, Inc.)

southern portion of the Ozarks were made of novaculite, a special type of sedimentary rock composed of microcrystalline quartz.

Stone tools made by early Native Americans included various sizes of projectile points, hand axes, blades for butchering animals, scrapers for working with hides, and grinding stones. Small projectile points (commonly referred to as "arrowheads") sometimes turn up in such places as plowed fields. Native Americans also fished using bone fishhooks and nets weighted down by stone sinkers. Natural occurring rock shelters are known to have been used by some of the earliest Native Americans (Fig. 2.5). The example shown is located on the Ozark Plateau in south central Missouri. The Rockhouse Cave site on the Buffalo River in Arkansas is another much larger example in the Ozarks. However, early Native Americans mostly constructed shelters out of the materials available to them, which included putting together a framework of poles cut from tree saplings and covered with large pieces of bark or animal hides.

2.7 The Woodland Period

The impact of Native Americans of the Appalachians and Ozarks on the landscape was likely very minimal until about 2600 years ago when what is known as the Woodland Period began. The beginning of this period marked a transition from a nomadic lifestyle to a more sedentary existence. Hunting and foraging for wild edible plants became confined to limited geographic areas and cultivation of domesticated crop expanded. Regional trade networks appeared, the technology involved in manufacturing stone tools became more developed, and ceramics began to be used for cooking vessels and other items. Towards the end of the Woodland Period about 1000 years ago, Native Americans were clearing small areas of forest for growing crops such as corn and squash. The cultivation of squash dates from about 2000 to 800 years ago, while domesticated corn followed about two centuries later.

By the middle of the Woodland Period, trade networks stretched from the east coast to the Rocky Mountains and from the Gulf of Mexico to the Great Lakes. As such, Native American societies in the Appalachians and Ozarks were linked (at least minimally) for the first time. Trade goods included a wide variety of artistic and exotic items, including mica, copper, obsidian, marine items, and high quality chert for tool making. During this period of time, Native Americans undoubtedly cleared some areas for their crops, and in at least some areas of the Appalachians they purposely set fires in at least some areas of forest. Nevertheless, their total impact on forests was likely minimal.

The late Woodland Period marked the beginning of a fully developed agricultural system and the appearance of large permanent villages, sometimes with protective palisades. Shelters for individual families or groups were well constructed. It has been suggested that individuals from these large villages dispersed during the winter in small bands to establish hunting camps, but these would have left no evidence of their existence. However, because of the larger more concentrated village populations, more forests had to be cleared away to make room for fields of cultivated crops. Forests were cleared away by girdling (cutting away a ring of bark) or setting fire to a group of trees. The ash from the burned trees was sometimes used as a fertilizer for their crops. Large villages were invariably located in valleys within both the Appalachians and Ozarks. Upland areas were more sparsely populated and were used mostly for hunting. It is likely that higher elevation areas in the Appalachians were visited only infrequently, since there were few edible plants other than blueberries and the animals present were mostly small.

The Woodland Period witnessed the most complete, complex, and well-developed indigenous culture ever to exist in the eastern and central United States. During the latter part of the period, there was some movement among the populations of Native Americans, although the reasons for these remain unclear. The biggest technological change was the gradual transition from the spear to the bow and arrow. Ceramics became more elaborate, and in some parts of the Appalachians and Ozarks, the Native Americans left evidence of elaborate burial ceremonies.

To what extent did Native Americans impact forests prior to their initial contact with Europeans? Experts have never been in complete agreement, but evidence from charcoal deposits and pollen records along with the very earliest accounts from Europeans suggest they deliberately set fire to forests to accomplish a number of things. First, as already mentioned, they used fire to clear land for growing squash, corn, and other domesticated crops. Second, they used fires to improve visibility within the forest, to facilitate travel, and to alter the forest habitat to make it more favorable for the growth of more desirable plants such as blueberries. Third, they used fire to keep areas already clear of forests open and more favorable for some of the animals that were hunted. A fourth use is more problematic to substantiate, but it is likely that they use fires to drive some of the larger animals they hunted, making them easier to kill.

Extensive fire histories have been developed for the Ozarks of Missouri and Arkansas. Prior to arrival of the Europeans, it was clear that the region consisted of forests with varying fire regimes. While evidence for fire was clear in the tree ring record as early as the late 1600s, the fires used by early settlers as they cleared land and became established were more frequent and often more intense than the "pre-settlement" fire regime (Guyette et al. 2002). Humans and fire are inextricably linked and even though the spread of fire may be controlled by topography (i.e., the forests on the most steep landscape do not easily burn), humans apparently can override this constraint with intentional ignitions.

2.8 European Contact

The arrival of the Europeans had a serious disruptive effect on indigenous Native Americans, with warfare and diseases ultimately taking a major toll on their populations (Yarnell 1998). Trade between the Native Americans and Europeans and hunting by the latter represented the basis of the initial economy in the two regions. However, by the late 1700s, European settlers and in some cases their African slaves had established a much more extensive economy based on farming, livestock, and various small-scale industries. Hunting considerably reduced the populations of some species of mammals, while grazing by livestock, logging, and clearing of land for agriculture greatly affected forest vegetation. These trends continued into the early 1800s, but two other things materialized that would greatly impact the Appalachians and to a lesser extent the Ozarks. The first of these was the first mining of coal in the two regions and the other was a major increase in logging operations. In addition, lead was mined in the northern portion of the Ozarks. All of these were well underway by the 1840s but were interrupted by the Civil War.

From 1861 to 1865, some portions of the Appalachians suffered greatly from the turmoil and violence of the Civil War. One of the major theaters of the war was the Shenandoah Valley. As a result of military operations, numerous farms were destroyed, pillaged or abandoned, large numbers of livestock killed, and large areas of forest cleared or burned (Stephenson 2013). The last of these is probably the least appreciated consequence of the Civil War. One report published in the late 1870s indicated that forests in the mountainous areas west of the Shenandoah Valley were reduced by more than a third between 1860 and 1870.

One little-known incident of the Civil War occurred in 1863, when fire escaped from the campfire of Confederate scouts in the eastern portion of the Allegheny Plateau in West Virginia. One of the fires swept through the eastern side of Allegheny Mountain in Pendleton County, as described in an early report. "Eyewitnesses have stated that the flames overtopped the tallest pines and advanced ten miles (16 km) an hour" (West Virginia Conservation Commission 1909). More often, fire was used to reduce forest that could conceal enemy troops or cannons. Nevertheless, large tracts of forest still existed in both the Appalachians and Ozarks in 1870. This was particularly in the latter region, where much of the logging occurred in the very late 1800s and early 1900s.

The Ozarks of Missouri and Arkansas witnessed numerous skirmishes during the Civil War, most of these undocumented but resulting in destruction of the landscape and depopulation of small communities. Pivotal battles such as those at Pea Ridge and Wilson's Creek remain historically prominent, but the entire cultural and physical landscape of the Ozark was transformed by the Civil War (Blevins 2019).

After the end of the Civil War, mining became increasingly important and logging expanded to more and more areas of forests. More land was cleared for agriculture and more trees were cut to supply fuel for heating, to power steam engines, and to make way for industrial production of an increasing number of different types of items to satisfy public demand. In some parts of the Appalachians, forests that occurred on the slopes of some mountains were often removed just to create pastures. The latter was especially true for some the high-elevation areas of the Allegheny Plateau covered by red spruce forests. Farmers would purchase large tracts of land, paying only 25–75 cents/acre (0.4 ha). In order to convert the forests to pasture for their cattle, they would use a process called "hacking." This involved girdling all of the trees, which would cause them to die. Later, the area would be burned to get rid of the dead wood and to kill the undergrowth. After a few years, enough native grasses would have come in to support a few cattle. Obviously, this was a terrible waste of forested land. This and other aspects of the forests of the Appalachians and Ozarks will be discussed in later chapters.

References

Blevins B (2019) A history of the Ozarks, The conflicted Ozarks, vol 2. University of Illinois Press, Champaign

Clark GM, Ciolkosz EJ (1988) Periglacial geomorphology of the Appalachian highlands and interior highlands south of the glacier border—a review. Geomorphology 1:191–220

Coope GR (1968) Insect remains from silts below tills at Garfield Heights, Ohio. Geol Soc Am Bull 79:753–756

Craig AJ (1969) Vegetational history of the Shenandoah Valley. Geol Soc Am Spec Pap 123:283–296

Delcourt PA, Delcourt HR (1981) Vegetation maps for eastern North America: 40,000 yr B.P. To the present. In: Romans RC (ed) Geobotany II. Plenum, New York, pp 123–165

Grayson DK (1991) Late Pleistocene mammalian extinctions in North America: taxonomy, chronology, and explanations. J World Prehist 5:193–231

Guilday JE (1982) Appalachia 11,000–12,000 years ago: a biological review. Archaeol East N Am 10:22–26

Guilday JE, Parmalee PW, Hamilton HW (1977) The Clark's cave bone deposit and the late Pleistocene paleoecology of the central Appalachian Mountains of Virginia. Bull Carnegie Mus Nat Hist 2:1–87

Guyette RP, Muzika R-M, Dey DC (2002) Dynamics of an anthropogenic fire regime. Ecosystems 5:472–486

Guyette RP, Stambaugh MC, Lupo A, Muzika R-M, Dey DC (2006) Oak growth in mid-western North America linked with post glacial climate epochs in the North Atlantic. Pages News 14(2):21–22

Hawksley O (1986) Remains of quaternary vertebrates from Ozark caves and other miscellaneous sites. MO Speleol Soc 26:1–67

Hinkle CR, McComb WC, Safley JM, Schmalzer PA (1993) Rock outcrop communities: limestone, sandstone, and granite. In: Martin WH, Boyce SG, Echternacht AC (eds) Biodiversity of the Southeastern United States: upland terrestrial communities. John Wiley and Sons, New York, pp 203–253

Jones RA, Williams JM, Jackson ST (2017) Vegetation history since the last glacial maximum in the Ozark highlands (USA): a new record from cupola pond, Missouri. Quat Sci Rev 170:174–187

King JE (1973) Late Pleistocene palynology and biogeography of the Western Missouri Ozarks. Ecol Monogr 4:539–565

Lesser WH (1993) Prehistoric human settlement in the upland forest region. In: Stephenson SL (ed) Upland forests of West Virginia. McClain Printing Company, Parsons, WV, pp 231–260

Owen W (2012) History of natural plant communities of the South. Vascular Plants and Paleobotany. Pennsylvania: US Department of Agriculture, Forest Service, Northeastern Research Station

Smith EN (1984) Late quaternary vegetational analysis of Cupola Pond, Ozark National Scenic Riverways, Southeastern Missouri MS Thesis. University of Tennessee, Knoxville

Stephenson SL (2013) A natural history of the Central Appalachians. West Virginia University Press, Morgantown

Wagner DP, McAvoy JM (2004) Pedoarchaeology of Cactus Hill, a sandy Paleoindian site in southeastern Virginia, U.S.A. Geoarchaeology 19(4):297–322

West Virginia Conservation Commission (1909) Report of the West Virginia Conservation Commission, 1908. The Tribune Printing Company, Charleston, WV

Whittaker RH (1956) Vegetation of the great smoky mountains. Ecol Monogr 26:1–80

White PS, Buckner ER, Pittillo JD, Cogbill CV (1993) High-elevation forests: northern hardwood forests and associated communities. In: Martin WH, Boyce SC, Echternacht AC (eds) Biodiversity of the Southeastern United States: upland terrestrial communmities. John Wiley and Sons, New York, pp 305–337

Yarnell SL (1998) The Southern Appalachians: a history of the landscape. U.S. Department of Agriculture, Forest Service, Northeastern Research Station, Pennsylvania

Chapter 3
Forests of the Appalachians

Abstract The pre-settlement forest vegetation of the Appalachians consisted of two broadly defined deciduous forest associations (oak-chestnut and mixed mesophytic) at lower and intermediate elevations and one subalpine coniferous association found at higher elevations. Because of the virtual elimination of American chestnut (*Castanea dentata*) by the chestnut blight fungus in the first half of the twentieth century, the oak-chestnut association is now best referred to as the mixed oak association. These forest associations occur in three physiographic provinces (Ridge and Valley, the Blue Ridge, and the Appalachian Plateaus) of the Appalachians. Various species of oak, especially northern red oak (*Quercus rubra*), chestnut oak (*Q. montana*), and white oak (*Q. alba*) along with several species of hickory (*Carya* spp.), are consistently present in the mixed oak association, and red spruce (*Picea rubens*) is the defining species of the subalpine coniferous association. The mixed mesophytic association is characterized by a diverse assemblage of trees, including sugar maple (*Acer saccharum*), American beech (*Fagus grandifolia*), black birch (*Betula lenta*), and Canadian hemlock (*Tsuga canadensis*). Each of these is considered in some detail in this chapter.

Keywords Arcto-Tertiary Geoflora · E. L. Braun · Forest associations · Mixed mesophytic forests · Oak-chestnut forests · Present-day forests · Subalpine coniferous forests

3.1 The Arcto-Tertiary Geoflora

Prior to the beginning of the Pleistocene about two and half million years ago, it is thought that what has been named the Arcto-Tertiary Geoflora existed as a continuous band across higher latitudes in the Northern Hemisphere. However, with the onset of the cooler temperatures associated with the continental glaciers of the Pleistocene, the assemblage of species making up the Arcto-Tertiary flora became fragmented. The existence of the Arcto-Tertiary Geoflora would account for the

S. L. Stephenson, R.-M. Muzika, *Forests of the Southern Appalachian Highlands and Interior Highlands of the United States*, Plant and Vegetation 20,
https://doi.org/10.1007/978-3-032-01155-8_3

disjunct distribution of a number of plants currently found in eastern North America and eastern Asia (Deng et al. 2015). Indeed, the forests of the two regions located thousands of kilometers apart are remarkably similar, with species of oak, maple, buckeye, chestnut, walnut, and magnolia found in both regions. This compositional similarity is also reflected in secondary strata of vegetation, and anyone walking through a forest in eastern China during the spring would recognize many of the spring ephemerals appearing on the forest floor. It was first thought that the same species were shared between the two regions, but it is now known that they are closely related species that are morphologically similar. This disjunct distribution pattern is not limited to vascular plants but also applies to some species of fungi.

Braun (1950) hypothesized that the mixed mesophytic forests of the southern Appalachians represented the direct descendent of the ancient Tertiary forest, and thus from it all of the other forest types were derived. However, she assumed that the area in which it occurred during the Pleistocene had remained relatively intact, and palaeobotanical data obtained since she proposed her hypothesis indicate that this was not the case.

Delcourt and Delcourt (1981) used fossil pollen data from 162 sites in eastern North America to examine the distribution of forest types and individual tree taxa over past 20,000 years. At the height of the Wisconsin glaciation, climatic conditions immediately south of the glacier were too severe for most species of trees to survive, with the notable exceptions (as noted earlier) of spruce and pine. However, with improving conditions after the end of the Pleistocene, trees began migrating northward. The rates of migration depended largely on the dispersal potential of their fruits or seeds. Trees such as oaks and hickories, with their large fruits, could not migrate as rapidly as those producing small, wind-dispersed fruits. Nevertheless, the northward migration of oak proceeded more rapidly than might have been expected. Oaks, which were mostly confined to areas of the Appalachians south of approximately 33 °N during the height of Wisconsin glaciation, had already extended to far north as about 47 °N by 8000 years ago. The role of animal vectors (e.g., squirrels and birds such as the blue jay [*Cyanocitta cristata*]) also cannot be discounted in the migration of trees, since they are capable of transporting seeds and fruits (even large examples) over short distances (Barnett 1977).

As noted earlier in this book, the higher temperatures and more xeric conditions associated with the Holocene climatic optimum affected the distribution patterns of individual species of trees and the overall pattern of vegetation across the landscape in both the Appalachians and Ozarks. However, by the end of the Holocene climatic optimum 5000 years ago, contemporary forests of the two region were largely in place.

3.2 Forest Associations

Braun (1950) recognized two broadly defined deciduous forest associations in the Appalachians. These were **oak-chestnut** (now mixed oak) and **mixed mesophytic**. In addition to these two deciduous forest association, a third type of association—**subalpine coniferous**—occurs at higher elevations throughout the region. Because of the virtual elimination of American chestnut (*Castanea dentata*) by the chestnut blight fungus, oak-chestnut is no longer appropriate as a name of this forest association, which is now more appropriately called mixed oak. However, because the primary objective of this chapter is to provide a description of pre-settlement vegetation in the Appalachians, Braun's designation of the forest association as oak-chestnut is retained just for the purposes of this book.

The forests of the Appalachians occupy a region of considerable size and environmental diversity with respect to land form, climate, soils, and geology. As mapped by Braun (1950), Küchler (1964), and Stephenson et al. (1993), oak-chestnut forests cover major portions of the Blue Ridge and Ridge and Valley physiographic provinces, while mixed mesophytic forests are characteristic of the Appalachian Plateaus physiographic province. Subalpine coniferous forests are found in the relatively limited areas of these three physiographic provinces where mountains reach elevations high enough to provide the cool, moist conditions required for this forest type to exist.

3.3 The Oak-Chestnut (Now Mixed Oak) Association

3.3.1 Forests of the Ridge and Valley

The pre-settlement forests of the northern Ridge and Valley section of the Appalachians were probably largely dominated by oak and American chestnut, since similar areas of western Maryland were described by Shreve et al. (1910) as "being made up predominantly of chestnut, chestnut oak (*Quercus montana*), and northern red oak (*Q. rubra*), which together form 75% to 90% of the stand [standing timber]." Early survey notes recorded by George Washington during the period of 1748–1752 in the general area of what is now Hardy County in West Virginia suggest an abundance of white oak (*Q. alba*), with northern red oak, chestnut oak, American chestnut, hickory, and pine the most numerous of the other trees present. Farther south, in southwestern Virginia, pre-settlement forests also seemed to have contained a high percentage of oaks.

There is little question that American chestnut was an important component of pre-blight forests throughout the Ridge and Valley, especially on the side slopes of ridges. In a survey carried out in the "mountain region of Virginia" (largely encompassing the Ridge and Valley physiographic province but also including parts of the Blue Ridge physiographic province), Lotti and Evans (1943) reported that oaks represented about two-thirds of all broadleaf trees, with chestnut oak, white oak, and northern red oak the most important species present. At the time their survey was carried out (in 1940), living American chestnut trees were no longer present, but the authors suggested that "chestnut was probably the predominant species of the old-growth (i.e., pre-settlement) forests)." In eastern Tennessee, early accounts of forest composition list American chestnut as one of the most important trees present (Foster and Ashe 1905; Hall 1910).

Fig. 3.1 Oak forest in the Ridge and Valley of southwestern Virginia. (Photo by Steven L. Stephenson)

In the Ridge and Valley, the oak-chestnut forest type was best expressed on open, south-facing slopes (Fig. 3.1). Prior to the arrival of the chestnut blight fungus, American chestnut sometimes made up 80% of the forest canopy in these situations (Braun 1950; Agrawal and Stephenson 1995). In late spring, when the white flowers of American chestnut were at their peak, the slopes looked as if snow had fallen. Although no longer present in the forest canopy, American chestnut still survives in the understory of the forests in which it was once dominant as sprouts from old stumps and root systems. These sprouts typically do not survive long enough to produce flowers, but on rare occasions, it is still possible to encounter a small American chestnut producing the characteristic white flowers. Non-flowering small sprouts remain relatively common in some forests in the Ridge and Valley. However, few sprouts have persisted in the most mesic habitats, where shading and competition are likely to have been detrimental to their survival. Stephenson et al. (1991) reported sprout densities as high as 2500/stems per hectare for some forests in southwestern Virginia. The demise of American chestnut in the eastern United States will be discussed in another chapter.

The post-blight forest vegetation of the Ridge and Valley has been the subject of a number of studies (e.g., McCormick and Platt 1980; Stephenson 1982; Stephenson 1986; Stephenson and Fortney 1998; Stephenson et al. 1991; Hart et al. 2008). Based on the results of these studies, the most consistently abundant species of trees now found on open slopes are chestnut oak, northern red oak, red maple (*Acer rubrum*), black oak (*Quercus velutina*), and white oak, with the first two species especially dominant. Pignut hickory (*Carya glabra*) is not uncommon and is a dominant species in some situations. On protected slopes and in mountain coves, other species such as sugar maple (*Acer saccharum*), basswood (*Tilia americana*), cucumber magnolia (*Magnolia acuminata*), black birch (*Betula lenta*), black cherry (*Prunus serotina*), white ash (*Fraxinus americana*), and yellow buckeye (*Aeseulus flava*) are characteristic of the forest canopy. American beech (*Fagus grandifolia*), Canadian hemlock (*Tsuga canadensis*), and yellow birch (*Betula alleghaniensis*) are important at some localities, especially those on north-facing slopes and at higher elevations (Table 3.1).

Table 3.1 Composition of representative mixed oak (formerly oak-chestnut) forests at six localities in the Ridge and Valley physiographic province

Taxon	1	2	3	4	5	6
Quercus montana	28	31	13	12	2	35
Q. rubra	17	29	18	13	5	5
Q. alba	13	4	8	3	33	8
Q. coccinea	10	<1	1	9	3	0
Acer rubrum	7	10	7	11	0	<1
Q. velutina	5	5	<1	1	7	6
Pinus rigida	5	<1	0	0	0	0
Carya spp.	4	8	21	1	13	16
Oxydendron arboreum	3	2	<1	4	0	0
Nyssa sylvatica	3	1	0	1	0	0
Betula lenta	<1	2	3	5	0	0
Robinia pseudoacacia	<1	3	2	4	0	<1
Magnolia acuminata	<1	3	<1	4	0	0
Liriodendron tulipifera	0	<1	<1	9	5	9
Acer saccharum	0	1	6	4	0	<0
Tsuga canadensis	0	<1	0	<1	0	0
Fraxinus americana	0	<1	3	<1	0	<0
Tilia americana	0	<1	1	<1	0	0
Other species	1	1	10	7	5	20

Data are important value indices based on 100. (Adapted from Stephenson et al. 1993)

Throughout the Ridge and Valley and Blue Ridge, forests dominated by northern red oak are locally extensive on higher ridges and summits (>900 m) (Fig. 3.2). For example, Ayres and Ashe (1905) indicated that "towards the mountain summits the red oak prevails and is often almost the only oak on the higher crests." Stephenson and Adams (1989) obtained data from 13 high elevation sites (>1100 m) in the Ridge and Valley and northern Blue Ridge. The majority of these sites were situated on sites with a northern exposure. Northern red oak was dominant in the canopy with a mean importance value of 70, with red maple, black birch, and white oak the most important associates. Downy serviceberry (*Amelanchier arborea*) and striped maple (*Acer pensylvanicum*) were present as small trees. The shrub stratum was dominated by ericaceous shrubs, including mountain cranberry (*Vaccinium erythrocarpum*), lowbush cranberry (*V. pallidum*), mountain laurel (*Kalmia latifolia*), flame azalea (*Rhododendron calendulaceum*), and minnie-bush (*R. pilosum*).

Forests dominated by northern red oak are not as characteristic of the Cumberland Plateau and Allegheny Plateau. At lower elevations than those at which red spruce (*Picea rubens*) is found, northern hardwood forests exist in these two regions. Northern hardwood forests are especially well developed on higher ridges of the southern Blue Ridge and the eastern portion of the Alleghany Plateau (Fig. 3.3). On the Allegheny Plateau, these forests generally occur at elevations above 915 m, but they sometimes extend downslope to elevations as low as 800 m. Among the more

Fig. 3.2 High elevation red oak forest on Locust Mountain at the southern edge of the Ridge and Valley in southwestern Virginia. (Photo by Gary P. Fleming)

Fig. 3.3 Northern hardwood forest on the floor of Canaan Valley in West Virginia. (Photo by Steven L. Stephenson)

characteristic trees associated with northern hardwood forests are sugar maple, American beech, yellow birch, hemlock, basswood, red maple, cucumber tree, yellow poplar (*Liriodendron tulipifera*), northern red oak, and chestnut oak (Core 1966). The first three species are the usual dominants on most sites, but it is not uncommon for one or more of the other species to occur as canopy co-dominants or dominants at a particular site.

Muzika et al. (1999) reported the results of a study in which 105 study plots were sampled on the Fernow Experimental Forest in West Virginia. The Fernow Experimental Forest is located within the Allegheny Mountain portion of the Allegheny Plateau, and most of the forests present would be classified as northern hardwood. A least some species of oaks were found in almost all study plots sites, along with American beech and black birch. Forests on mesic north-facing slopes usually contained some combination of northern red oak, sugar maple, tulip tree, black cherry, white ash, basswood, Canadian hemlock, and cucumber tree. Hemlock was sometimes the overwhelming dominant on poorly drained sites or along streams. On drier south and east-facing slopes, northern red oak, white oak, chestnut oak, scarlet oak, red maple, red maple, black gum, sassafras, and sourwood were the primary canopy trees. Black locust and Fraser magnolia (*Magnolia fraseri*) were occasionally minor components of the forest on dry and moist sites, respectively (Stephenson et al. 1993).

Other species that occasionally occur in northern hardwood forests include several species of hickory, black locust (*Robinia pseudoacacia*), eastern white pine (*Pinus strobus*), hercules' club (*Aralia spinosa*), sourwood (*Oxydendrum arboretum*), and striped maple (*Acer pensylvanicum*). The last species is relatively common as a shrub or small tree in northern hardwood forests, but it only rarely attains tree size (i.e., having a trunk diameter of at least 10 cm).

In the southern Blue Ridge, northern hardwood forests occur at appreciably higher elevations (usually >1300 m) due to the effect of decreasing latitude. The assemblage of trees present is similar to that of characteristic of the Allegheny Plateau (White et al. 1993). Biodiversity and total cover of the herbaceous stratum in northern hardwood forests are as high as in many examples of mixed mesophytic forests in the Southern Blue Ridge.

The northeast-southeast orientation of the ridges of the Ridge and Valley physiographic province create a situation in which slope exposure and topography become factors of some significance (Stephenson and Mills 1999). Slopes with a southern and western exposure are warmer and drier than slopes with a northern exposure. These environmental conditions are sufficient to affect the distribution of trees. For example, chestnut oak is likely to be more important on slopes with a southern exposure than on slopes with a northern exposure, while just the reverse is true for northern red oak. There have been several studies that have examined the differences that exist between the two different exposures (e.g., Shanks and Norris 1950; Mowbray and Oosting 1968; Stephenson 1982). In the latter study, weekly air temperatures at 1 m and soil temperatures at a depth of 10 cm were taken for an entire field season. The mean air temperature (25.2 °C) recorded for the southern exposure was appreciably higher than the mean temperature (23.9 °C) recorded for the northern exposure, while the values for soil were 15.7 °C (south) and 13.2 °C (north). These data provide some evidence of the magnitude of the differences in microclimate that exist for the two types of exposures.

The ridges of the Ridge and Valley are capped with resistant sandstone or quartzite. The less resistant shales and limestones have eroded away, producing the intervening valleys. Soils derived from sandstone or quartzite are sandy, nutrient poor, and acidic (Stephenson et al. 1986), while those derived from limestone and shales are more loamy, fertile, and much less acidic. These differences are reflected to a considerable extent in the composition of the forests present. For example, such trees as sugar maple, chinquapin oak (*Quercus muehlenbergii*), and eastern redbud (*Cercis canadensis*) are typically associated with limestone-derived soils. In relatively open forests located at elevations below 750 m, flowering dogwood (*Cornus florida*) commonly occurs on such soils, although this species also can survive on sites with sandstone-derived soils.

Most forests in the entire Appalachian region have been logged or otherwise disturbed by human activities, and very few old-growth remnants remain. The average age of forests in the region is typically <100 years, which means that most have gone through at least two cutting cycles. The impact of the chestnut blight fungus on these forests will be discussed in detail in another chapter.

The Shenandoah Valley (also called the Valley of Virginia) is the easternmost valley of the Ridge and Valley physiographic province and is bordered on the east by the Blue Ridge physiographic province. The forests of the Shenandoah Valley have a long history of human disturbance dating back to the early 1700s. Because the original forests were disturbed so long ago, there are almost no data on their composition. The original forests were likely similar to those of the smaller valleys to the west, with oaks among the more important canopy components. White oak was the most important oak, with chinquapin oak, various hickories, and red cedar (*Juniperus virginiana*) also present, especially over open areas of the valley floor. In more sheltered locations and along streams, the forests were more mixed mesophytic in character. Unlike the forests of the Ridge and Valley to the west, the shrub stratum was not dominated by ericaceous taxa. There appears to be little evidence to suggest that American chestnut was ever relatively common throughout the Shenandoah Valley (Fleming 2012).

Red pine (*Pinus resinosa*) is a characteristic tree of the Hemlock-White Pine-Northern Hardwoods Forest Region (*sensu* Braun 1950) of northeastern North. The natural range of this species does not extend into the Appalachians except for two indigenous populations in the Ridge and Valley physiographic province in eastern central West Virginia. These two populations, one located in Hardy County and the other in Pendleton County, are disjunct from the closest indigenous populations in northcentral Pennsylvania by more than 200 km. Both populations occur at an elevation >900 m on steep slopes with a northern or northwest exposure. Red pine was clearly dominant (mean IV = 83.0) when the forests at these two sites were sampled in the mid-1980s (Stephenson et al. 1986). The only other species of trees shared by the two sites were black birch, pitch pine (*Pinus rigida*), and northern red oak. Because red pine has been widely used for reforestation throughout the Allegheny Plateau, it is now found at number other localities.

Northern white-cedar (*Thuja occidentalis*) is another tree found in the Appalachians but not in the Ozarks. The natural range of this species encompasses southern Canada and the northern United States, but scattered populations occur as far south as eastern Tennessee. There are historical records of indigenous populations in western North Carolina, but these no longer exist. Northern white-cedar is associated with calcareous soils, but it can be found in sites ranging from low wet areas (swamps) to limestone bluffs. A description of the vegetation of the first type was provided by Ogle (1989) for a site (known as Barns Chapel Swamp) in the Ridge and Valley of Washington County in southwestern Virginia. Northern white cedar occurred on slightly elevated areas of the swamp, along with oaks, hickories, buckeye, and America holly (*Ilex opaca*). Adams and Stephenson (1989a) sampled forests dominated by northern white cedar at two localities in the Ridge and Valley of West Virginia. Both occurred on rather steep slopes with a northwestern exposure in areas underlain by limestone. At the first locality, where the forest was located on the lower slope of ravine, white walnut (*Juglans cinerea*) was the second ranking species. At the second locality, northern white cedar occurred on an open rocky slope, and red cedar was the other more important species present. One remarkable feature of northern white cedar is its potential longevity. Some individuals have been documented to be more than 1000 years old!

3.3.2 Forests of the Blue Ridge

The forests of the Blue Ridge were the first forests in the Appalachian region to be encountered by Europeans. As a result, they have a very long history of disturbance and descriptions of pre-settlement vegetations are practically nonexistent (Stephenson et al. 1993). In what is perhaps the earliest report of forests in just the Blue Ridge (Hough 1878), the Blue Ridge was described as "mostly covered with forests of white, black, red and rock (*Quercus montana*) oak, hickory, chestnut, locust (*Robinia pseudoacacia*), birch, some excellent yellow pine (probably *Pinus echinata*), and other trees."

Based on a fairly comprehensive survey of the mountainous regions of southwestern Virginia (including portions of both the Blue Ridge and Ridge and Valley) carried out in 1900 and 1901, Ayres and Ashe (1905) indicated that various species of oak, along with American chestnut, where the most consistently abundant trees present. In that portion of the Blue Ridge that encompasses the Balsam Mountains and adjacent areas in extreme southwestern Virginia, oaks (45%), American chestnut (20%), Canadian hemlock (4%), birch (3%), maple (3%), and basswood (3%) were the major trees present. At the time this survey was carried out, "forests of large areas" were already limited to higher elevations in the region.

As noted earlier in this book, the Blue Ridge physiographic province consists of two rather distinct sections separated by the Roanoke River in southwestern Virginia. The southern section of the Blue Ridge occupies a much larger area, extends over a wider range of elevation, and is more varied with respect to both topography, geology, and vegetation. This topographic diversity and wide range of elevation allows species associated with different habitats to exist in a spatially small areas. The southern section is more floristically diverse than the northern section and is characterized by a higher concentration of endemic plants.

Some of the trees found in the forests of the southern Blue Ridge prior to logging could reach a considerable size. The very tallest was yellow poplar, (also known as tulip tree) which on favorable sites could exceed a height of 50 m, with a trunk diameter of 1.8 m. American chestnut (height of 40 m and a trunk diameter of 3.0 m) was another very tall tree. Both white oak and northern red oak could reach at least 30 m, with a trunk diameter of 1.8 m. However, much shorter individuals of white oak could have larger trunks. The largest conifer was eastern white pine, which could exceed 45 m and have a trunk 1.6 m in diameter. In subalpine coniferous forests, the very largest red spruce trees could exceed 40 m, with a trunk 1.0 m in diameter. Trees this size typically occurred in protected ravines. On exposed ridgetops and upper slopes, red spruce trees were much smaller.

In the northern section, forest composition varies significantly with geological substrate and associated soils. On the western flank of the province, which is underlain by quartzites and other metasedimentary rocks, mixed oak forests with chestnut oak, northern red oak, and white oak are the primary species that occupy the relatively xeric higher crests of the main ridges. Mountain laurel is often conspicuous in the understory. Johnson and Ware (1982) described the forests of the central Blue Ridge of Virginia. They indicated that chestnut oak was more important than northern red oak at lower elevations, while the reverse was true at higher elevations. In more mesic situations such as on protected slopes and in ravines, Canadian hemlock, eastern white pine (*Pinus strobus*), and birches are components of the canopy. Black birch is consistently present, but yellow birch is largely limited to the highest elevations and north-facing slopes. Side slopes with deeper soils are mostly covered with oak-hickory forests, with pignut hickory, red hickory (*Carya ovalis*), shagbark hickory (*Carya ovata*), and bitternut hickory (*C. cordiformis*) the most common species of hickory. As a general observation, hickories appear to represent a more important component of the canopy than in the forests of most areas of the Ridge and Valley to the west. Yellow poplar is consistently present and sometimes dominant in lower elevation post-cultivation areas (Table 3.2).

In the most xeric situations, especially on slopes with a southern exposure, there are some areas of pine forests dominated by such species as Virginia pine (*Pinus virginiana*), with Table Mountain pine and pitch common also common in some localities (Fig. 3.4). At the very lowest elevations, tree diversity increases as other species of trees make their appearance in the tree stratum. In most respects, the forests on this part of the northern Blue Ridge resemble those of the Ridge and Valley at the same latitude.

Although many soils of the northern Blue Ridge are not unlike those commonly found throughout the Appalachians, the eastern portions of the range are underlain by very ancient pre-Cambrian rocks that include metabasalt (greenstone) and a complex of base-rich granites and granitic gneisses. These have weathered away to produce unusually fertile soils with high levels of calcium and magnesium. Such soils are favorable for the development of diverse oak-hickory forests with a lush ground cover of nutrient-demanding herbaceous plants such as black cohosh (*Actaea racemosa*) and horsebalm (*Collinsonia canadensis*) not typically found in most Appalachian ridge forests. The coves in these areas are especially rich, with sugar maple and basswood usually among the overstory dominants. Interestingly, in the portion of the Blue Ridge where the underlying rocks are pre-Cambrian in age, the ridges are generally less steep and have broader crests than elsewhere in this physiographic province. The soils on the crests are deeper and less rocky than the soils on the side slopes of the ridges. This is reflected in the composition of the vegetation, with relatively more mesic oak and oak-hickory present on the crests and more xeric often oak-dominated forests on the side slopes.

The southern section of the Blue Ridge broadens into a gently rolling plateau up to 100 km wide in some places and with a few peaks exceeding 1800 m. The northernmost mountains in this section of the Blue Ridge that reach a considerable elevation are the Balsam Mountains in southwestern Virginia, with Mt. Rogers at 1746 m.

Table 3.2 Composition of representative mixed oak (formerly oak-chestnut) forests at six localities in the Blue Ridge province

Taxon	1	2	3	4	5	6
Quercus rubra	22	31	25	36	2	35
Acer saccharum	19	<1	<1	5	<1	0
Q. montana	15	14	21	11	60	0
Betula lenta	11	8	9	0	0	5
Tsuga canadensis	11	0	<1	0	<1	4
Q. alba	10	16	2	0	0	0
Carya spp.	3	19	17	12	0	0
Fraxinus americana	2	3	<1	4	0	0
Tilia heterophylla	2	1	1	2	0	5
Acer rubrum	1	1	5	15	11	9
Liriodendron tulipifera	1	0	2	0	1	0
Robinia pseudoacacia	1	4	2	6	1	2
Nyssa sylvatica	<1	0	1	0	3	0
Q. coccinea	0	0	1	0	7	0
Q. velutina	0	1	<1	0	1	0
Oxydendron arboreum	0	0	0	0	6	0
Magnolia acuminata	0	0	<1	2	0	0
Pinus rigida	0	0	0	0	3	0
Other species	1	1	10	7	5	20

Data are important value indices based on 100. (Adapted from Stephenson et al. 1993)

Fig. 3.4 Pine forest in the Ridge and Valley of southwestern Virginia. (Photo by Gary P. Fleming)

These mountains should not to be confused with mountains with the same name located in western North Carolina. The vegetation of the southern section of the Blue Ridge is best exemplified by the vegetation of the Great Smoky Mountains as described by Whittaker (1956). The field studies that served as the basis for this publication were carried out during the 1947 field season, and there were still some highly diseased but still living American chestnut trees in some of the forests he sampled. Whittaker recognized a number of different forest types, and those with

Fig. 3.5 Mixed mesophytic cove forest in the Grayson Highlands of the southern Blue Ridge in southwestern Virginia. (Photo by Gary P. Fleming)

the greatest areal extent on xeric sites associated with slopes with a southern exposure were pine-oak at lower elevations, oak-hickory at intermediate elevations, and northern hardwoods at higher elevations but below the elevations at which subalpine coniferous forests occurred. On more mesic slopes with a northern exposure, successional forests occupied the most extensive areas at lower elevations and mixed mesophytic forests occurred at higher elevations (Fig. 3.5). The latter were best developed in mountain valleys (coves). In addition to high tree diversity, these coves provide the habitats for numerous spring ephemerals, whose flowers display an array of vibrant colors and forms across the forest floor just before the leaves become fully developed on the trees overhead.

At higher elevations in the southern Blue Ridge, just below or sometimes within the lowest portion of elevation zone occupied by red spruce or red spruce/Fraser fir forests, American beech-dominated forests occur. These can be found on broad open ridges or concave areas of upper slopes. The former are referred to as beech orchards, whereas the latter are known as beech gaps. Sometimes, these beech gaps exist as isolated islands in what are otherwise coniferous forests (White et al. 1993). Adams et al. (2003) described similar beech-dominated forests for study sites located on the Allegheny Plateau, the Ridge and Valley, the Southern Blue Ridge, and the Balsam Mountains portion of the Blue Ridge in southwestern Virginia (Fig. 3.6). In each of these four regions, beech shared dominance with a different compliment of species, with northern red oak the second leading dominant in two regions, sugar maple in one, and yellow birch/mountain silverbell (*Halesia tetraptera*) in one (Table 3.3).

American beech-dominated forests are found at much lower elevations in the Ozarks and typically occur in ravines. Red maple is often the second leading dominant. In general, the understory layer of beech-dominated forests is much less developed in the Ozarks than in the Appalachians, presumably because of the generally drier conditions. As will be mentioned later in this book, published data on the composition of beech forests have considerable historical value because of the devastation caused by beech bark disease.

Fig. 3.6 Beech forest in the Ridge and Valley of southwestern Virginia. (Photo by Steven L. Stephenson)

Table 3.3 Composition of beech forests at four localities in the Appalachians

Taxon	1	2	3	4
Fagus grandifolia	43	47	33	34
Halesia tetraptera	12	0	0	0
Betula alleghaniensis	11	9	6	4
Tsuga canadensis	11	0	1	<1
Aesculus flava	10	2	0	0
Acer saccharum	3	24	8	0
A. rubrum	3	2	10	5
Tilia americana	3	2	2	0
Quercus rubra	<1	<1	15	35
Betula lenta	<1	<1	3	7
Q. alba	0	0	1	1
Prunus serotina	0	4	8	0
Liridodendron tulipifera	0	0	6	0
Other species	5	8	6	13

Data are important value indices based on 100. (Adapted from Adams et al. 2003)

In the southern Blue Ridge below the beech forest zone or sometimes replacing it in the same elevation zone are northern hardwood forests. These are especially well developed on north-facing slopes. In these forests, as already noted, the dominant species include such examples as sugar maple, yellow birch, hemlock, and northern red oak. With decreasing elevation, northern hardwood forests transition into cove forests where these exist. Northern hardwood forests are especially common in the Allegheny Mountain section of the Alleghany Plateau. In the Ozarks, the northern hardwood forest as traditionally defined does not exist.

Successional forests, with Virginia pine a common and consistent component, were described by Whittaker as occupying xeric sites at lower elevations and on open ridgetops in the Great Smoky Mountains National Park. This is still the case today. However, on slightly more mesic sites, especially those that were once non-forested, yellow poplar is often the primary dominant.

Carolina hemlock (*Tsuga carolinana*) is an Appalachian endemic that occurs in the Blue Ridge physiographic province from northeastern Georgia to southwestern Virginia. The species typically occurs on xeric exposed slopes with acidic and nutrient-poor soils. Carolina hemlock is not abundant, and sizeable indigenous populations exist at relatively few localities. Populations of this uncommon species, like those of the more common Canadian hemlock, have been devastated by the hemlock woolly adelgid. At one such locality (Bottom Creek Gorge in southwestern Virginia), sampled in 1997, Carolina hemlock had an IV of 49, with chestnut oak (19), black oak (12), and red maple (12) the more important associated species (Rentch et al. 2000). Carolina hemlock is grown as a cultivated ornamental and thus now occurs in the Appalachians beyond the Blue Ridge.

Whittaker used a gradient analysis approach to describe the patterns of vegetation across the Great Smoky Mountains. Gradient analysis seeks to relate changes in forest composition to the differences that exist along some environmental factor that can be identified in nature. Whittaker included different forest types as well as areas of non-forest vegetation such as grassy balds and heath balds in his analyses. He also considered the secondary strata of vegetation (i.e., shrubs and herbaceous plants). He was able to relate changes in vegetation to the environmental complex-gradients associated with differences in elevation and site moisture conditions. Whittaker used the term "complex-gradient" because each of these two gradients actually involves a number of interacting factors. For example, elevation alone simply serves as quantifiable measure when defining gradients. However, the actual factors responsible for changes in vegetation include such things as the general trends of decreasing temperature and increasing moisture as elevation increases in a mountainous region. The gradient analysis approach to describing patterns of vegetation has been used in numerous subsequent studies carried out in various other places in both the United States and other places in the world.

The sampling of forest vegetation by Whittaker included only a portion of the park, but in the mid-1930s, when the Great Smoky Mountains National Park was being established, Frank Miller was hired to do a comprehensive survey of the entire park (MacKenzie and White 1998). Miller recorded a total of 55 species of trees and small trees, a figure that clearly indicated the very high diversity of the forests of the Southern Appalachians. Interestingly, the total number of trees and small trees now known from the region is more than 100, which is about equal to the number known from all of Europe! At the time the sampling by Miller was carried out, which was more than a decade after the chestnut blight fungus had reached the Southern Appalachians, American chestnut remained the dominant species throughout the park. Miller listed the other more widespread and important species as Canadian hemlock, red spruce, yellow birch, chestnut oak, northern red oak, and American beech.

3.4 The Mixed Mesophytic Association

Mixed mesophytic forests are the predominant forest type throughout those portion of the Appalachians (essentially the Cumberland Plateau and the Allegheny Plateau) that fall within the Mixed Mesophytic Association as delimited by Braun (1950). However, examples of this forest type also commonly occur in the Blue Ridge and Ridge and Valley where they are found in environmentally favorable (i.e., moist) topographic situations such as on north-facing slopes and in ravines. Diversity of the tree stratum is not as high in the northern Blue Ridge, Ridge and Valley, and northern Allegheny is not as high as in the Cumberland Plateau and southern Blue Ridge, but representatives of the same "core" group of tree species are present in every instance. These include species such as sugar maple, American beech, yellow birch, Canadian hemlock, basswood, yellow poplar, and cucumber magnolia (Table 3.4).

3.4.1 Forests of the Cumberland Plateau

The topography of the Cumberland Plateau, like that of the Allegheny Plateau, is generally flat to rolling but extensively dissected, especially towards its western edge. The rugged Cumberland Mountains, essentially an extension of the Allegheny Mountains to the northeast, form the eastern boundary of the Cumberland Plateau. The Allegheny Plateau is located to the northeast, while the southern and western

Table 3.4 Composition of representative mixed mesophytic forests at three localities in the Blue Ridge (BR) and three localities in the Ridge and Valley (RV) physiographic provinces

Taxon	BR			RV		
Liriodendron tulipifera	14	0	0	4	3	12
Carya spp.	12	1	0	31	16	15
Tsuga canadensis	12	<1	12	<1	0	0
Betula alleghaniensis	10	6	4	0	<1	0
Acer saccharum	8	28	27	8	11	11
Quercus rubra	6	4	4	8	6	1
Acer rubrum	6	4	0	2	1	7
Betula lenta	6	2	0	<1	0	0
Fraxinus americana	4	12	1	2	2	4
Tilia heterophylla	3	22	19	11	5	<1
Quercus montana	3	<1	0	5	6	0
Fagus grandifolia	2	14	9	<1	22	16
Prunus serotina	2	<1	2	0	<1	0
Magnolia acuminata	<1	1	0	2	0	<1
Aesculus flava	0	2	14	1	5	2
Halesia carolina	0	0	6	0	0	0
Other species	11	2	2	26	23	30

Data are important value indices based on 100. (Adapted from Stephenson et al. 1993)

borders transition into the Oak-Hickory-Pine Forest Region and Western Mesophytic Forest Regions, respectively.

Early studies of the forests of the Cumberland Plateau include those of Foley (1901, 1904) and Hall (1910). These were followed by the extensive and detailed descriptions provided by Braun in her book *Deciduous Forests of Eastern North America* (1950). All three authors were fortunate enough to be able to observe the forests of the Cumberland Plateau prior to extensive logging. All of them recognized a series of general forest types. More xeric sites such as ridgetops and upper slopes were dominated by American chestnut, chestnut oak, black oak, scarlet oak (*Quercus coccinea*), and various hickories. On lower side slopes and in more mesic situations, other species entered the canopy. The most diverse forests occurred in coves, where more than thirty species could be recorded at a single locality. However, species such as American beech, yellow poplar, basswood, sugar maple, Canadian hemlock, white oak, American chestnut, buckeye, mountain silverbell, and northern red oak were the usual dominants. Other species such as yellow birch, black birch, black cherry, cucumber tree, white ash, red maple, back gum, black walnut, and several species of hickory were consistently present but only rarely dominant.

The mixed mesophytic forests of the Cumberland Plateau are considered to be the most biodiverse temperate deciduous forests anywhere in the world. As noted earlier, Braun (1950) considered these forests as direct descendants of the ancient Tertiary forest and from which the other forest types were derived. The latter is no longer considered to be the case, but the very high biodiversity of forests of the Cumberland Plateau cannot be questioned. Descriptions and characteristics of old-growth mixed mesophytic forests in this portion of the Appalachians were provided by Martin (1975, 1992).

With increasing latitude in eastern North America, the general pattern is for biodiversity of the tree stratum to decrease. When the lower latitudinal limit for coniferous forests is reached in the northern United States and extreme southern Canada, biodiversity of trees at a particular locality may drop to only two or three species. Also, as latitude increases and/or elevation decreases, many of the same species characteristically associated with northern hardwood forests of the Appalachians become the most common species present.

3.4.2 Forests of the Allegheny Plateau

There are few descriptions of pre-settlement forests of the Allegheny Plateau. Fontaine (1876) described the forests of the region as "composed mainly of white, chestnut, black, and red oaks; chestnut, hickory, (yellow) poplar, ash (presumably white ash), sugar maple, (Canadian) hemlock, beech, locust, and black walnut (*Juglans nigra*)." Of these, white oak seems to have been the most important species. Somewhat later, Brooks (1911) indicated that the forests were made of "many species of hardwoods." The latter is an appropriate description, since both the pre-settlement forests and the relatively little disturbed forests that still exist are characterized by a high diversity of tree species. These forests occur across the landscape as a mosaic of different combinations of the same species mentioned by Fontaine along with a few other species (e.g., red maple and black cherry) that he did not mention.

American chestnut clearly occurred in the pre-settlement forests, although this species was not nearly as abundant as it was in the Ridge and Valley. Chestnut Ridge, a geological feature extending south from Pennsylvania to the city of Morgantown in northern West Virginia, was named on the basis of the occurrence of American chestnut. Forests of the western portion of the Allegheny Plateau are less diverse than those of the eastern portion, although except for the highest elevations, the same assemblage of species is involved. These forests, which Core (1966) referred to as the xeric expression of mixed hardwood forests, consist predominately of several species of oak, with various hickories, black gum (*Nyssa sylvatica*), sourwood, red maple, Virginia pine, and shortleaf pine (*Pinus echinata*) also present. As the name of this forest type implies, this these forests occur on relatively dry sites.

The mesic expression of mixed hardwood forests (sensu Core) is found on slopes with a northern exposure or on the lower side slopes of valleys. As many as 25 species of trees can be found in this type of forest, although not all at one locality. These include basswood, slippery elm (*Ulmus rubra*), yellow poplar, cucumber tree, American beech, sweet (black) birch, and Canadian hemlock. The number of species of trees is comparable to what would be expected for a mixed mesophytic forest as described by Braun (1950).

The Appalachian Plateaus physiographic province is underlain by sedimentary rocks such as sandstones, conglomerates, and shales, with limestones exposed at the surface in some places. In general, these rocks have produced soils that are acidic and nutrient poor. Of considerable importance to the recent history of the region is the fact that these layers of sedimentary rocks have interbedded coal seams. From the mid 1700s onward, removal of coal has had serious disruptive consequences for much of the region.

3.5 The Subalpine Coniferous Forest Association

A half century or so ago, the term "boreal forests" sometimes was used for what are more appropriately called subalpine coniferous forests. This is not surprising, since the two types of forests are similar in appearance and are usually dominated by spruce. However, the species of spruce involved are different, with red spruce characteristic of the high elevations in the Appalachians but not in boreal forests. Instead, white spruce (*Picea glauca*) and black spruce (*Picea mariana*) are the two species of spruce in boreal forests. Another difference is that paper birch (*Betula cordifolia* and *B. papyrifera*) replace yellow birch as the "birch component" of boreal forests. Paper birch does occur in the southern portion of the Appalachians but is quite rare.

A transition occurs between boreal forests and subalpine coniferous forests, with the latter no longer subalpine in location, in southern Canada. Balsam fir (*Abies balsamea*), which occurs along with red spruce in some areas of the Appalachians, does continue northward into boreal forests and thus is one thing the two types of forests do share in common.

Subalpine coniferous forests are overwhelming dominated by red spruce, but in a few places red spruce shares dominance with fir. Yellow birch and Canadian hemlock are the two other types of trees most likely to be present (Table 3.5). In the southern portion of the Appalachians, the fir is the narrow endemic Fraser fir (*Abies fraseri*), but in the northern portion, it is balsam fir. Balsam fir commonly occurs along with red spruce in the Appalachians north of the region considered in this book but reaches its southernmost limit in the mountains of eastern central West Virginia and northern Virginia (Stephenson and Adams 1986). Indigenous populations of balsam fir occur at just four localities in West Virginia and at only two of these (Blister Run and Canaan Valley) does the species exist in reasonable numbers. Small populations also occur at a two localities (Hawksbill Mountain and Stony Man Mountain in Shenandoah National Park) in the Blue Ridge in northern Virginia.

Subalpine coniferous forests occupy a much smaller area of the Appalachians than oak-chestnut forests and mixed mesophytic forests. Various estimates have been generated, and these range from about 70 km^2 to a much as 300 km^2 (White et al. 1993). This wide range of values reflects the difficulty in delimiting the boundary between subalpine coniferous forests and the other forests that surround them. This is especially true at the lower elevation limits for subalpine coniferous forest where spruce can occur along with broadleaf trees, either as scattered individual trees or as interspersed pockets of red spruce in an otherwise broadleaf forest. The latter is especially true in on the Allegheny Plateau, where it is the product of

Table 3.5 Composition of representative subalpine coniferous forests at six localities in the Appalachians. Data are important value indices based on 100

Taxon	1	2	3	3	5	6
Picea rubens	49	46	42	21	20	24
Betula alleghaniensis	32	24	12	1	11	28
Abies fraseri/A. balsamea	6	3	43	72	46	0
Tsuga canadensis	5	<1	0	0	13	3
Amelanchier arborea	3	<1	0	0	2	1
Acer pensylvanicum	1	<1	<1	0	1	3
A. spicatum	1	1	2	0	0	4
A. saccharum	1	3	0	0	<1	1
A. rubrum	<1	2	0	0	5	11
Sorbus americana	<1	2	1	5	0	1
Fagus grandifolia	1	11	0	0	<1	11
Prunus pensylvanica	<1	1	<1	0	1	0
Aesculus flava	1	<1	<1	0	0	0
Other species	<1	4	0	0	1	8

Adapted from White et al. (1993)

irregular patterns of succession following extensive logging (Clarkson 1964). For example, in a study of post-logging red spruce forests on the Allegheny Plateau carried out in 1977, red spruce had an importance value (hereafter abbreviated as IV) of only 30 in canopy of the eight sites sampled. Yellow birch (IV = 24) and red maple (IV = 18) were the only other trees with an IV greater than ten. The fact that the forests in which red spruce occurs can vary considerably in composition is reflected in the fact that 14 other species of trees were recorded from these eight sites.

Regardless of the figure used for the extent of the subalpine coniferous forests, they occupy only a small percentage of the total area of the Appalachians. This percentage would have been higher prior to the extensive logging of red spruce that took place in the late 1800s and early 1900s. For example, on the Allegheny Plateau, it has been estimated that red spruce was reduced by about 90% as a result of logging by the late 1890s (Millspaugh 1891). In more southern portions of the Appalachians, the most intensive period of logging occurred somewhat later in the 1910s and 1920s. As will be described later in this book, logging operations were often accompanied by fire. The two types of disturbance—logging and fire—created habitats that were favorable for certain species of small trees not typically found in undisturbed forests. These were fire cherry (*Prunus pensylvanica*) and mountain maple (*Acer spicatum*). The latter is a species usually associated with rocky areas.

The lower elevation limit for subalpine coniferous forests in the Southern Appalachians is approximatively 1350 m. However, with increasing latitude, this forest type is found at lower elevations. In central Virginia and West Virginia, subalpine coniferous forests can occur as low as 975 m on mountain ridges and summits (White et al. 1993). In West Virginia, red spruce can occur at even lower elevations in bogs and other poorly drained areas (Core 1966). Balsam fir is found as low as 990 m in Canaan Valley, where it occurs along with red spruce in poorly drained areas. Yellow birch and Canadian hemlock are present along with balsam fir and red spruce on slightly more elevated areas in the valley. The only other locality in West Virginia (Blister Run) where balsam fir occurs in reasonable numbers also is a poorly drained area but at a higher elevation (1120 m) near the top of a mountain (Fig. 3.7).

Some parts of the southern Blue Ridge were never subjected to logging, including areas within the Great Smoky Mountains National Park. These old-growth forests give some indication of what the subalpine coniferous forests of the Allegheny Plateau must have looked prior to logging. There are very few localities on the Alleghany Plateau or in the northern Ridge and Valley where red spruce trees escaped logging. Adams and Stephenson (1989b) identified four such sites, although in only one (the Gaudineer Scenic Area in West Virginia) were there more than a few spruce trees present. Based on data from increment growth cores extracted from some of the larger red spruce trees at each site, all four sites had trees >250 years old, with the maximum age 368 years.

Fig. 3.7 Balsam fir forest at Blister Run in the Allegheny mountains of West Virginia. (Photo by Steven L. Stephenson)

Fig. 3.8 Fraser fir forest on Mt. Rogers in southwestern Virginia. (Photo by Gary P. Fleming)

The most extensive subalpine coniferous forests with or without Fraser fir present occur in the southern Blue Ridge. The northernmost locality for Fraser fir is Mt. Rogers (1746 m) in southwestern Virginia, where this type of forest is found on the summit at elevations above about 1700 m (Fig. 3.8). Balsam fir (IV = 72) is the most abundant species present, with spruce (IV = 21) much less important. The only other species present are yellow birch and mountain ash (*Sorbus americana*) (Stephenson and Adams 1984). Fraser fir is lacking but well-developed red spruce forests are present on Whitetop Mountain (elevation 1684 m), located a relatively short distance away from Mt. Rogers (Fig. 3.9).

Ten areas in the southern Blue Ridge surpass 1680 m in elevation. Seven of these have well-developed subalpine coniferous forests containing both red spruce and Fraser fir. One has red spruce but no Fraser fir, while two others have only scattered red spruce or fir trees present (White et al. 1993). Although broadly similar in composition, the subalpine coniferous forests of the southern Blue Ridge tend to exhibit greater growth and thus are taller and have a greater biomass.

One other noticeable difference is that these forests also have a greater cover of herbaceous vascular plants (especially ferns) and bryophytes on the forest floor. Some of these are endemic to subalpine coniferous forests.

Fig. 3.9 Red spruce forest on Whitetop Mountain in southwestern Virginia. (Photo by Steven L. Stephenson)

Soils associated with subalpine coniferous forests are thin, rocky (often exceedingly so), and nutrient-poor. In this respect they are similar to the soils associated with most other forests in the Appalachians. However, they differ in one major respect. This is the presence of a thick layer of slowly decomposing leaf litter (mostly from the dominant needle-leaved conifers) over the soil surface. This litter layer obscures the fact that there is little actual soil as the latter is usually defined. Soils are acidic, often with a pH lower than 4.0 (Adams and Stephenson 1989b). When this forest type occurs in the muck soils of wetland situations, the pH can be very low (occasionally <3.0).

References

Adams HS, Stephenson SL (1989a) Distribution and ecology of northern white cedar (*Thuja occidentalis* L.) in West Virginia. Va J Sci 40:62

Adams HS, Stephenson SL (1989b) Old-growth spruce stands in the mid-Appalachians. Vegetatio 85:45–56

Adams HS, Stephenson SL, Ware S, Schnittler M (2003) Forests of the central and southern Appalachians and eastern Virginia having beech as a major component. Castanea 68:222–231

Agrawal A, Stephenson SL (1995) Recent successional changes in a former chestnut-dominated forest in southwestern Virginia. Castanea 60:107–113

Ayres HB, Ashe WW (1905) The southern Appalachian forests, Professional Paper No. 37. U.S. Geological Survey, Washington, DC

Barnett RJ (1977) The effect of burial by squirrels on germination and survival of oak and hickory nuts. Am Midl Nat 98:39–330

Braun EL (1950) Deciduous Forests of Eastern North America. Blakiston Company, Philadelphia

Brooks AB (1911) Forestry and Wood Industries, West Virginia geological survey, vol 5. ACME Publishing Company, Morgantown, WV

Clarkson RB (1964) Tumult on the mountains. McClain Printing Company, Parsons, WV

Core EL (1966) Vegetation of West Virginia. McClain Printing Company, Parsons, WV

Delcourt PA, Delcourt HR (1981) Vegetation maps for eastern North America: 40,000 yr B.P. to the present. In: Romans RC (ed) Geobotany II. Plenum, New York, pp 123–165

Deng T, Nie Z-L, Drew BT, Volis S, Kim C, Xiang C-L, Zhang J-W, Wang Y-H, Sun H (2015) Does the Arcto-tertiary biogeographic hypothesis explain the disjunct distribution of northern hemisphere herbaceous plants? The case of *Meehania* (Lamiaceae). PLoS One 10(2):e0117171. https://doi.org/10.1371/journal.pone.0117171
Fleming GP (2012) The nature of the Virginia flora. In: Weakley AS, Ludwig JC, Townsend JF, Crowder B (eds) Foundation of the Flora of Virginia project Inc., Richmond. Botanical Research Institute of Texas Press, Fort Worth, pp 24–75
Foley J (1901) A working plan for southern hardwoods, and its results. Yearbook of the Department of Agriculture, Washington, DC
Foley J (1904) Conservative limbering at Sewanee, Tennessee, United States Department of Agriculture Forest Bulletin No. 39. United States Department of Agriculture Forest, Washington, DC
Fontaine WW (1876) Timber—its distribution and development. In: Maury MF, Fontaine WW (eds) Resources of West Virginia. The Register Company, Wheeling, WV, pp 142–161
Foster HD, Ashe WW (1905) Chestnut oak in the southern Appalachians, USDA Forest Service circular no. 135. Government Printing Office, Washington, DC
Hall RC (1910) Preliminary study of forest conditions in Tennessee. Tenn Geol Surv Ser Bull 10:1–56
Hart JL, van de Gever SL, Grissino-Mayer HD (2008) Forest dynamics in a natural area of the southern ridge and valley, Tennessee. Nat Areas J 28:275–289
Hough FB (1878) Report upon forestry prepared under the direction of the commissioner of agriculture in pursuance of an act of congress approved august 15, 1876. Government Printing Office, Washington, DC
Johnson GG, Ware S (1982) Post-chestnut forests in the central blue ridge of Virginia. Castanea 47:329–343
Küchler AW (1964) Potential natural vegetation of the conterminous United States, American Geographical Society special publication no. 36. American Geographical Society, New York
Lotti T, Evans TC (1943) The forest situation in the mountain region of Virginia, USDA Forest Service, Forest Service survey release no. 15. Appalachian Forest Experiment Station, Ashville, NC
MacKenzie MD, White PS (1998) Vegetation of great smoky mountains national park. Castanea 63:323–336
Martin WH (1975) The Lilley Cornett woods: a stable mixed mesophytic forest in Kentucky. Bot Gaz 136:171–183
Martin WH (1992) Characteristics of old-growth mixed mesophytic forests. Nat Areas J 12:127–135
McCormick JF, Platt RB (1980) Recovery of an Appalachian forest following the chestnut blight or Katherine Keever—you were right! Am Midl Nat 104:264–273
Millspaugh CF (1891) The black spruce. W V Agr St Rp 3:98–101
Mowbray TB, Oosting HJ (1968) Vegetation gradients in relation to environment and phenology in a southern blue ridge gorge. Ecol Monogr 38:309–344
Muzika R-M, Stephenson SL, Adams HS, Lawrence DW, Miller GW (1999) Patterns of composition in the tree stratum of the Fernow Experimental Forest and adjacent portions of the Otter Creek Wilderness Area. In: Eckerlin RP (ed), Proceedings of the Appalachian Biogeography Symposium, Virginia Museum of Natural History Special Publication Number 7. pp 35–44
Ogle D (1989) Burns chapel swamp: an unusual arbor-vitae (*Thuja occidentalis*) site in Washington County, Virginia. Castanea 54:200–202
Rentch JS, Adams HS, Coxe RB, Stephenson SL (2000) An ecological study of a Carolina hemlock (*Tsuga caroliniana*) community in southwestern Virginia. Castanea 65:1–8
Shanks RE, Norris FH (1950) Microclimatic variation in a small valley in eastern Tennessee. Ecology 31:532–539
Shreve F, Chrysler MA, Blodgett FH, Besley FW (1910) The plant life of Maryland, Maryland Weather Service, Special Publication No. 3. John's Hopkins Press, Baltimore

Stephenson SL (1982) Exposure-induced differences in the vegetation, soils, and microclimate of north- and south-facing slopes in southwestern Virginia. VA J Sci 33:36–50
Stephenson SL (1986) Changes in a former chestnut-dominated forest after a half century of succession. Am Midl Nat 116:173–179
Stephenson SL, Adams HS (1984) The spruce-fir forest on the summit of mount rogers in southwestern Virginia. Bull Torrey Bot Club 111:69–75
Stephenson SL, Adams HS (1986) An ecological study of balsam fir communities in West Virginia. Bull Torrey Bot Club 113:372–381
Stephenson SL, Adams HS (1989) Th high-elevation red oak (*Quercus rubra*) community type in the mid-Appalachians. Castanea 54:217–229
Stephenson SL, Fortney RH (1998) Changes in forest composition on the southwest-facing slope of beanfield mountain in southwestern Virginia. Castanea 63:482–488
Stephenson SL, Mills HH (1999) Contrasting vegetation of noses and hollows in the valley and ridge province, southwestern Virginia. Bull Torrey Bot Soc 126:197–212
Stephenson SL, Adams HS, Lipford ML (1986) Ecological composition of indigenous stands of red pine (*Pinus resinosa*) in West Virginia. Castanea 51:341–341
Stephenson SL, Adams HS, Lipford ML (1991) The present distribution of chestnut in the upland forests of the mid-Appalachians. Bull Torrey Bot Club 118:24–32
Stephenson SL, Ash AN, Stauffer DF (1993) Appalachian oak forests. In: Martin WH, Boyce SC, Echternacht AC (eds) Biodiversity of the Southeastern United States: upland terrestrial communities. John Wiley and Sons, New York, pp 255–303
White PS, Buckner ER, Pittillo JD, Cogbill CV (1993) High-elevation forests: northern hardwoods forests and associated communities. In: Martin WH, Boyce SC, Echternacht AC (eds) Biodiversity of the Southeastern United States: upland terrestrial communities. John Wiley and Sons, New York, pp 305–337
Whittaker RH (1956) Vegetation of the great smoky mountains. Ecol Monogr 26:1–80

Chapter 4
Forests of the Ozarks

Abstract Forests of the Ozarks largely contain elements of the deciduous forest farther east but transition into the great plains to the west and the southern pine forests to the south. The deciduous forests of the Ozarks are largely oak dominated, but within some areas there is a strong component of shortleaf pine. The latter species was more common in the presettlement forest than its current distribution. Although the topography is not dramatic and elevational differences are modest, local topographic relief plays an important primary role in controlling the composition of the vegetation and affects the disturbance regime because of the exposure of geologic parent material at different elevations and the importance of slope position and aspect. Southward into the Ouachita Mountains, shortleaf pine becomes increasing dominant. Restoration of pine forests and fire mediated disturbance regimes have become important throughout the Ozarks, especially in the northern portion of this physiographic province.

Keywords Fire · Forest-grassland ecotone · Forest restoration · Oaks · Presettlement vegetation · Shortleaf pine

4.1 Introduction

In her book *Deciduous Forests of Eastern North America*, Braun (1950) considered the Interior Highland (herein referred to as the Ozark) forests to represent the best and most concentrated expression of the Oak-Hickory Forest Region. While that may be accurate, her comment somewhat discounts the complexity of an area where there is no question that both oaks and hickories are abundant, but referring to the area as oak-hickory belies the complexity of forest types, natural communities, and landscape complexes found in within it. To avoid confusion with the other extensions of the Oak-Hickory Forest Region that extend into Wisconsin, southern Michigan, stretch into Canada, and reach as far south as Texas, the uniqueness and the character of the Ozarks must be carefully examined. The southern division of

S. L. Stephenson, R.-M. Muzika, *Forests of the Southern Appalachian Highlands and Interior Highlands of the United States*, Plant and Vegetation 20,
https://doi.org/10.1007/978-3-032-01155-8_4

the Oak-Hickory Forest Region, represented by the forests of the Ozark Plateaus province and Ouachita province, claim an identity of their own.

While hickories and particularly oak may be widespread and omnipresent in the Ozark Plateaus province and the Ouachita province, there are much higher concentrations of oak and more species of oak associated with particular environmental conditions. Moreover, the varying importance of shortleaf pine (*Pinus echinata*) throughout the region both in the past and at present more fully characterizes the heterogeneity of the landscape and complexity of forest community types.

A further challenge when considering the forest region of the Ozarks is that forests can range from being dense closed canopy forests to moderately open woodlands and even areas of very open savanna with few trees (Fig. 4.1). The transitions of forest with prairie to the north and west and with pure pine stands to the south create both ill-defined boundaries as well as a gradient of transitional forest types.

The region referred to herein as the Ozarks is characterized by considerable geologic, topographic, and edaphic complexity. These abiotic factors strongly contribute to the composition and structure of forests in the region and also present challenges in characterizing the forest vegetation. Generally speaking, the Ozarks might be considered as a large transitional area, occupying the vast ecotone that occurs between the great plains to the west and the deciduous forests to the east. Migration of particular species of trees over time, changes that have taken place in climate, and alteration of the landscape by humans all contribute to the complexity of forests—and indeed non-forest vegetation—in this transitional area. Given the geographical position of the Ozarks in the lower Midwest, the region has experienced incursion and retreat of species from the forests in the east and south and from grasslands to the north and west. Just as global climatic patterns, particularly those related to temperature and soil moisture content, dictate where on the planet forests might occur, precipitation and soil type strongly dictate the location of forested areas in the Ozarks. Furthermore, these climate and soil patterns also control forest productivity.

Fig. 4.1 Woodland in fall color within the Stegall Mountain glade-woodland-forest complex in the Ozark National Scenic Riverways, Missouri. (Photo by Susan Farrington)

4.2 The Ozarks

In the national hierarchical framework of ecological classification, the differentiating criteria for ecological sections are precipitation, temperature, geomorphology, major soil groups, and potential vegetation formations. The Ozarks represent an ecological section that is characterized by highly variable topography, ranging from relatively level plains to steep slopes. Soils are derived from a variety of geologic parent materials, commonly including dolomitic limestone and sandstones in the Ozarks Plateaus province and quartz, sandstones, and chert in the Ouachita province (Bryant et al. 1993).

4.3 General Characteristics of the Vegetation

As the westernmost extension of the eastern deciduous forest, the Ozarks share many characteristics of the forests farther to the east, but yet share some climatic, soil, and vegetation characteristic of the adjacent western areas, specifically the tall grass prairie. The ecological transitions between vegetation types are rarely abrupt and discrete. Therefore, boundaries and delineations are not only irregular, they also are subject to change. In the Ozarks, this results in forest communities that share some herbaceous component of prairies. The result is a relatively patchy community when viewed at a large scale. The patchiness and interdigitation of forest/grassland may be among the most defining attributes of the Ozark Plateaus province portion of the Ozarks. Farther south in the Ouachita province, the defining attribute might be the increasing abundance of pine.

Most of the forests of the Ozarks fall within what Braun (1950) referred to as the southern division of the Oak-Hickory Forest Region. Currently, these forests might be better described as oak-pine, although hickory is certainly present. In contrast with the other portions of the Oak-Hickory Forest Region that exist throughout the eastern United States, the Ozarks support a greater number of species of oak than anywhere else, with an appreciable number of species of oaks with distributions centered in the south. It is not uncommon to find black oak (*Quercus velutina*) or scarlet oak (*Q. coccinea*) in an Ozark forest along with southern red oak (*Q. falcata*). Less common species in the Ozarks might include Shumard oak (*Q. shumardii*) or bur oak (*Q. macrocarpa*).

4.4 Regional Differences

The Ozarks have the greatest extent of glade communities (open woodlands) in North America, and within the region there is a gradual increase in in pine forests from north to south. Although the most dramatic differences in forests occur between the Ozarks Plateaus province and the Ouachita province, there are notable

Fig. 4.2 Beech forest in the Boston Mountains of northern Arkansas. (Photo by Steven L. Stephenson)

difference within the former. The more northern portion of the province in Missouri tends to be less rugged topographically and have more xeric environmental conditions, whereas the rugged Boston Mountains in the southern portion of province in Arkansas support some plant communities and species that are characteristic of more mesic sites and thus more similar to the forests of the Appalachians (Stephenson et al. 2007). Notable examples are the beech-maple forests which are found in some areas of the Boston Mountains (Fig. 4.2). Beech (*Fagus grandifolia*) is absent from the Ozark Plateau of Missouri.

4.5 Presettlement Vegetation

Early explorers such as DeSoto and Coronado in the 1500s and a century later French explores such as LaSalle, all provided narratives to describe the vegetation of the Ozarks and what they encountered at that time. Much of the interpretation of the landscape suggests more open rather than densely forested areas (Beilmann and Brenner 1951). Henry Rowe Schoolcraft travelled throughout much of the Missouri and Arkansas Ozarks during the early 1800s. Although ostensibly the intent of his travel was to evaluate the area for mining (which he did), he published a written account of the area ("Journal of a Tour into the Interior of Missouri and Arkansas") that included descriptions of the natural landscape (Schoolcraft 1821). The descriptions suggest a forested landscape with frequent "openings" and also a patterned and heterogenous landscape. The differences in landscape as he described were as dramatic as moving through a barren landscape during part of the day and coming upon lofty pines later that day. The journals describe the vast difference in vegetation from the upland (often xeric) to the very dense vegetation associated with the Current, Gasoncade, Big Piney, and White Rivers. The general aspects of the natural landscape as well as the culture and history of the Ozarks have been synthesized and captured brilliantly by many writers, with Milton D. Rafferty and Brooks Blevins among the most noted.

In a more quantitative sense, notes from Public Land Survey records provide a general picture of the pre-settlement forest vegetation of Ozarks. There were about 285,000 trees surveyed between 1815 and 1850 in the Missouri portion of the Ozarks based on data from the Missouri General Land Office (GLO) dataset (J. Harlan, Geographic Resources Center, www.msdis.missouri.edu). In the Arkansas Ozarks, surveys began around 1820. From this information, a general and somewhat quantitative sense of pre-settlement forests and other vegetation can be derived from survey descriptions (Tucker 1990). The general concensus developed from of the vast number of publications examining the GLO notes described heterogenous areas ranging from pure stands of a single species to mixed species forests to open areas.

Presettlement records strongly support the co-occurrence of prairie, grassland types of communities interspersed with forests in the Ozark region. Moreover, numerous glades (referred to as barrens) occurred in the survey notes in the Ozarks, underscoring the significant of various communities such as woodlands, glades, and forests collectively forming a heterogenous landscape. The theme of variation in plant communities was recognized by Palmer (1921), who described the flora of the Ozarks and pointed out the occurrence of areas of exposed rock, relatively devoid of vegetation, using the term "barrens" to describe these. Palmer also noted the advancement of forests on previously non-forested landscapes.

4.6 Woodlands and Glades

Open woodland and glade communities account for a large area in the Ozarks and are often scattered throughout otherwise closed canopy forests (Fig. 4.3). The historical distribution of open woodlands and glades suggest that these occurred topographically on the plains or small hills as opposed to the steeper slopes and lower slope positions where closed canopy forests were found. As mentioned in other chapters, glades are often associated with shallow soil or even possibly occur on bare rock and support unique plant communities. Geological parent material may control the both the extent and the type of vegetation and can include trees despite being a general open landscape. For example, igneous glades may include short leaf pine or oak, while dolomitic glades often have scattering of chinquapin oak (*Quercus muehlenbergii*) or blackjack oak (*Quercus marilandica*).

4.7 Shortleaf Pine Communities

While oaks in general may dominate throughout the Ozarks, many areas of forest include pine as a codominant. In some places pure stands of shortleaf pine occur. In both the rugged topography of the river-associated hills of the Ozarks and in the more level or rolling upland plains or low hills, pine may be common as well. In the latter areas, pine may occur in woodlands or savannas owing to the frequency of

Fig. 4.3 Purplish-pink flowers of blazing star in the foreground of Coppermine Hollow Glade in the Angeline Conservation Area, Missouri. (Photo by Susan Farrington)

Fig. 4.4 Remnant older shortleaf pine (*Pinus echinata*) forest in the Missouri Ozarks. A developing understory including younger oak trees is evident. (Photo by Rose-Marie Muzika)

fires (Guyette et al. 2002). In the more topographically complex landscapes of the Ozarks, where fire might be less frequent than in the plains, pine is associated with acidic soils. Early records indicate a dominance of shortleaf pine in much of the Ozark Plateaus province, but current vegetation belies the previous dominance because until the most few recent few decades pine remained scattered and only a few remnant late successional and old growth pine forests existed (Fig. 4.4). Estimates suggest that pine-oak woodlands once covered an estimated 65% of the landscape of the Ozarks, and in Missouri alone covered 6.6 million acres (26,700 km^2).

In the Ozarks of Missouri and Arkansas, shortleaf pine tends to occur on drier slopes (e.g., south- and west-facing) and on sandstone derived soils. Topography plays an important role in that some of the best developed shortleaf pine forests occur at upper slope positions (coincidence with sandstone residuum) and on the steepest portions of the landscape such along the Elk River and White River near the Arkansas/Missouri border and in the Current River, Gasconade Rivers, and Meramec Rivers of the eastern central portion of the Ozark Plateau.

Pine forests were extensive and ranged from open pine woodlands to pure pine forests to pine-oak forests (Flatley et al. 2023). Fire played an important role in the distribution of pine and in the structure of the forest. Fire history reconstruction has revealed that fire historically was frequent across all pine-dominated forest types. Indigenous burning contributed to pine being maintained in Ozark forests, but natural fires though lightning strikes undoubtably occurred. Intensive logging of pines from 1880 to 1920 resulted in extreme depletion of the pine resource. Some of the largest pine sawmills in the United States were located in the Ozarks and accounted for removal of tens of thousands of acres of pine forest. Logging substantially reduced pine abundance and very frequent fires before, during, and following logging further eliminated advanced pine regeneration (Guyette et al. 2007), leaving a pine deficit. Fire suppression during the 1900s contributed to a reduction in pine regeneration, principally because fire reduced competition from hardwoods. Species of oak generally were able to capitalize on reduction of fire. A similar situation in the Ouachitas resulted in former pine dominated forests shifting into hardwoods.

Active restoration of pine forests is underway throughout the Ozarks. In part, this stems from the dramatic shifts in pine abundance that have occurred over the past 150 years and also recognition of the ecological significance of the communities associated with short leaf pine. Pine forests in the region provide habitat for many birds of conservation concern, notably the red-cockaded woodpecker (*Leuconotopicus borealis*), Bachman's sparrow (*Peucaea aestivalis*), prairie warbler (*Setophaga discolor*) among others. Fire regimes that promote pine also contribute to creating a diverse vegetation community, and ongoing restoration efforts in the Ozarks have enhanced plant community diversity following years of restoration with fire (Maginel et al. 2019), with a focus on pine restoration and pine-oak woodland restoration (Fig. 4.5).

Natural communities that have some shortleaf pine in the Ozarks include shortleaf pine-oak chert woodlands, shortleaf pine oak sandstone woodlands, shortleaf pine, and oak sandstone igneous woodlands. Species of blueberry (*Vaccinium*) and or warm season grasses such as big bluestem (*Andropogon gerardii*) often occur in association with pine and oak on these woodlands.

Fig. 4.5 A pine restoration site on the Mark Twain National Forest. The pine may have been seeded or planted in the 1950s and was overstocked. A timber sale occurred in 2013–2016 followed by at least three prescribed fires to stimulate the understory, enhance biodiversity, and reduce hardwood encroachment. (Photo by Denise Vaughn)

Farther south, in the east-west trending Ouachita Mountains, pine-oak and oak-hickory forests and woodlands occur in repeating east-west bands on the slopes of the ridges. As with the Ozarks, forests dominated (sometime exclusively) by short-leaf pine occur on areas where fire propagation occurred easily, such as on high plains and in areas with a level topography.

Shortleaf pine occurs on the west-facing and southern slopes, and, as in the Ozarks, in presettlement forests, shortleaf pine was no doubt quite dominant. The shale and sandstone dominated substrates in the Ouachitas support more widespread distribution of shortleaf pine in contrast to the limestones and dolomites of much of the Ozarks.

Development and maintenance of shortleaf pine (as with many pines, particularly those in the "hard pine" subgenus) requires fire. Fire in the presettlement Ouachita province was most frequent during the early eighteenth and mid nineteenth centuries. Over time, with extensive logging and subsequent replanting of loblolly pines, much of the shortleaf pine landscape in the Ouachita Mountains had been lost. Large scale restoration of the shortleaf pine-blue stem communities of the Ouachita Mountains has been underway since the mid-1990s. Thinning or removal of the overstory followed by prescribed burning are typical approaches used for restoration.

4.8 Oak-Dominated Communities

In his seminal publication *Spring Flora of* Missouri, Steyermark (1940) described the vegetation of the Missouri Ozarks by first identifying five "associations." These were sugar maple-bitternut hickory, sugar maple-white oak, oak-hickory, oak-pine, and white oak-red maple. These represented what might be considered the late successional forests, or in the terminology of early twentieth century plant ecologists, the "climax" forests. Steyermark generally considered the Ozark forests as mesophytic forests, a categorization that might be contrary to contemporary ecological thought. In her *Deciduous Forests of Eastern North America*, Braun (1950) described the mesophytic forest as a species rich forest of the Appalachians, while the Ozark Plateaus and Ouachita provinces were included in the southern division of the Oak-Hickory Forest Region, with only small pockets of forests dominated with mesophytic species in the Boston Mountains of the Arkansas Ozarks. These scientific distinctions and interpretation underscore the importance of time when considering forests, since change is inevitable with all natural systems over time. More recent research reveals a morenuanced if not reductionist understanding of Ozark forests. For example, Nelson (2010) described 15 forest communities or types less defined by species and more a function of parent material, soil type, and landscape position but associated with distinct species, and often several species. From the outset, plant ecologists have recognized the importance of geologic parent material and soil in the character of Ozark forest vegetation, particularly soil alkalinity and acidity.

With no fewer than 25 species of oak native to the forests of the Ozarks, this genus remains ubiquitous across the region and dominant in many forest types with some species occurring in non-forested plant communities. The vast range in the distribution of some species of oaks and their ecological amplitude (i.e., the type and range of local site conditions to which a particular species is best suited) control both the fine scale and broad scale importance of oak throughout the region. As an example, white oak (*Quercus alba*), which occurs across the entire eastern deciduous forest biome, achieves its highest levels of dominance and concentration in some areas of the Ozarks.

White oak distribution generally reflects broad climatic conditions, and among oaks, white oak is relatively tolerant of drier conditions. Although this species is widely found throughout the entire region, within a given forest in the eastern United States, it is not typically the dominant tree present. In contrast, white oak can be found as a dominant tree in many oak forests in the Ozarks, and there are notable old-growth forests which may be considered as the most mesic upland forests in the Ozarks despite the drought tolerance of white oak (Fig. 4.6). In contrast, the southern Ozarks of northern Arkansas and the Ouachita include more northern red oak (*Q. rubra*), a slightly more mesic species, more common here than in Ozarks of Missouri.

As a transitional zone, the Ozarks are characterized by numerous species that are found throughout adjacent biogeographical regions. Examples include oaks found in the Ouachitas and southern Ozarks such as Nuttall's oak (*Q. texana*), a species with a distinctly more southern distribution.

Similarly, southern red oak reaches its northernmost distribution in the Ozarks. A widespread species associated with the transitional area of the great plains and eastern deciduous forest, bur oak can be found in the Ozarks and Ouachitas, but is not particularly common. It has a bimodal distribution in that it is associated with both dry uplands as well as bottomlands and river floodplain forests. Shumard oak is sparsely distributed but reaches the northernmost distributional limit in the Ozarks.

Fig. 4.6 An old-growth forest in the Current River Hills of the Missouri Ozarks dominated by white oak (*Quercus alba*), some of which may be more than 300 years old. (Photo by Rose-Marie Muzika)

Post oak (*Q. stellata*) achieves its highest importance in eastern Texas, east Oklahoma, and in the Ozarks of Missouri and Arkansas. Post oak woodlands and sandstone glades can be found in the southern Ozarks of Missouri and the northern Ozarks of Arkansas. Interestingly, scattered old-growth individuals of post oak can be found in the forests of northern Arkansas (Bragg et al. 2012) and some of these trees are two to three centuries old. Community types that incorporate the driest conditions include blackjack oak (*Q. marilandica*) and post oak. These oaks are also found on glades and woodlands. Chinkapin oak (*Q. muehlenbergii*) has a distribution that is strongly midwestern and also tends to be an oak found on drier sites. It may be found on dolomitic glades along with eastern red cedar (*Juniperus virginiana*) or in dolomitic woodlands where the latter species may be among the oldest trees (occasionally >400 years) present since they grow slowly and persistently on low productivity sites. Of particular note is the maple-leaf oak (*Quercus acerifolia*). This species occurs on only four mountain ridges in Arkansas and is endemic to the Ouachita Mountains in west-central Arkansas.

In general, any combination of oak and other species may be found in an array of community types from woodlands to forests. The relative abundance of tree species may be related to substrate type and aspect. As general examples, the white oak-black oak dry mesic chert woodland community type may be found on north- or east-facing slopes while the mixed oak-hickory-dogwood community type occurs on igneous substrates.

Moving toward less xeric conditions, the communities dominated by oak may include black oak and scarlet oak in more closed canopy forests rather than in woodlands. Black oak is widely distributed but has often has a high importance in the Ozarks. Scarlet oak, which occurs widely in the Appalachians, is found nearly exclusively in the eastern Ozarks of Missouri. It is moderately drought tolerant but not tolerant of fire. Both black and scarlet oaks are dominant in many Ozark forests that had previously supported shortleaf pine. Following logging, the loss of the shortleaf pine seed source along with a reduction in fire frequency, black and scarlet oak became dominant on low productivity sites. During the latter half of the twentieth century, vast areas in the Ozarks were subject to oak decline, largely characterized by black and scarlet oak dieback and mortality. Oak decline has been a prevalent concern for at least four or five decades, and many factors associated with it, including low site productivity, altered disturbance regimes, and biotic agents such as armillaria root disease and the wood-boring red oak borer (*Enaphalodes rufulus*). In addition, drought as an inciting factor affects species such as scarlet and black, which have shorter lifespans than many other species of oak.

Resampling of early twentieth century inventories reveals that both shortleaf pine and species of oak have declined in abundance over time in the Arkansas Ozarks (Soucy et al. 2005; Chapman et al. 2006). Increases in species that are understory tolerant but fire sensitive reflect changing disturbance regimes when compared with the major events, namely logging and intense fires, that brought the dominance of pine and oak.

In summary, the forest communities of the Ozarks do not include the types found at higher elevations and the more mesic sites characteristic of the Appalachians.

However, except for those species limited to these two situations, the forests of the two regions share many features in common. Various species of oak, with most occurring in both regions, are typically the dominant trees present in most forest communities. In both regions, compositional differences in forest communities often can be related to differences in soil parent material and topography. Forests in both regions have been subjected to considerable disturbance in the past, which has greatly affected their present composition, especially in some localities. However, in both the Appalachians and Ozarks forests remain as a resource whose value should not be underestimated.

References

Beilmann AP, Brenner LG (1951) The changing forest flora of the Ozarks. Ann Missouri Bot Gard 38:283–291

Bragg DC, Stahle DW, Cerny KC (2012) Structural attributes of two old-growth cross timbers stands in western Arkansas. Am Midl Nat 157:40–55

Braun EL (1950) Deciduous forests of eastern North America. Blakiston Company, Philadelphia

Bryant WS, McComb WC, Fralish JS (1993) Oak-hickory forests (western mesophytic/oak-hickory forests). In: Martin WH, Boyce SC, Echternacht AC (eds) Biodiversity of the southeastern United States: upland terrestrial communities. John Wiley and Sons, New York, pp 143–201

Chapman RA, Heitzman E, Shelton MG (2006) Long-term changes in forest structure and species composition of an upland oak forest in Arkansas. For Ecol Manag 236:85–92

Flatley WT, Bragg LM, Bragg DC (2023) Dynamic fire regimes and forest conditions across three centuries in a shortleaf pine-oak forest in the Ouachita Mountains, Arkansas, USA. Ann Am Assoc Geogr 113(6):1365–1382. https://doi.org/10.1080/24694452.2023.2189937

Guyette RP, Muzika R-M, Dey DC (2002) Dynamics of an anthropogenic fire regime. Ecosystems 5:472–486

Guyette RP, Muzika R-M, Voelker SL (2007) The historical ecology of fire, climate, and the decline of shortleaf pine in the Missouri Ozarks. In: Shortleaf pine restoration and ecology in the Ozarks: proceedings of a symposium, General technical report NRS-P-15. US Department of Agriculture Forest Service, Northern Research Station, Newtown Square, PA, pp 8–18

Maginel CJ, Knapp BO, Kabrick JM, Muzika R-M (2019) Landscape-and site-level responses of woody structure and ground flora to repeated prescribed fire in the Missouri Ozarks. Can J For Res 48(8):1004–1014

Nelson PW (2010) The terrestrial natural communities of Missouri. The Missouri Department of Natural Resources

Palmer EJ (1921) The forest flora of the Ozark region. J Arnold Arbor 2(4):216–232

Schoolcraft HR (1821) Journal of a tour into the interior of Missouri and Arkansas from Potosi, or mine a Buron, in Missouri territory, in a south-west direction, toward the Rocky Mountains, performed in the years 1818 and 1819. Sir Richard Phillips, London

Soucy RD, Heitzman E, Spetich MA (2005) The establishment and development of oak forests in the Ozark Mountains of Arkansas. Can J For Res 35:1790–1797

Stephenson SL, Adams HS, Huebner CD (2007) Upland forest vegetation of the Ozark Mountains in northwestern Arkansas. Rhodora 109:197–221

Steyermark J (1940) Spring Flora of Missouri. Missouri Botanical Garden, St. Louis, MO

Tucker GE (1990) Pre-settlement vegetation of the Ozark National Forest. In: Restoration of old growth forests in the interior highlands of Arkansas and Oklahoma: proceedings of the conference, Winrock international, Morrilton, AR, September 19–20, 1990, p 67

Chapter 5
Beneath the Canopy in Forests of the Appalachians and Ozarks

Abstract Although forests are defined and described on the basis of the species of trees present in the canopy layer (or stratum), several other strata exist beneath the canopy in relatively undisturbed deciduous forests in the Appalachians and Ozarks. These other strata are recognized on the basis of the species, age, growth form, and size of the plants that make them up. For example, this is the case for the shrub and herbaceous strata, which are floristically very different. Trees temporarily share living space in some of these strata as they develop from seeding to mature individuals. The lowermost stratum (the ground layer), unlike the other strata, is dominated by non-vascular plants and lichens. In this chapter, the composition of each of these non-canopy strata is discussed, including the similarities and differences in the two regions. More detailed information is provided for some of the plants mentioned.

Keywords Groups of plants · Ground layer · Vascular plants · Non-vascular plants · Regional differences · Shade tolerance · Strata of vegetation · Vegetation composition · Vegetation structure

5.1 Introduction

Forests are defined and described on the basis of the species of trees present, but a forest community has a number of other floristic and structural components that are not quite so obvious. Forests (as is the case for any plant community) can be characterized on the basis of their floristic composition and physical structure. The latter refers to the sizes and general appearance (or physiognomy) of the various plants that make up the community whereas the former refers to the assemblage of taxa (usually but not always identified to level of species) present.

S. L. Stephenson, R.-M. Muzika, *Forests of the Southern Appalachian Highlands and Interior Highlands of the United States*, Plant and Vegetation 20,
https://doi.org/10.1007/978-3-032-01155-8_5

5.2 Structure of the Vegetation

On the basis of structure alone, three major layers (or strata) can be recognized within relatively undisturbed deciduous forests in the Appalachians or Ozarks. These three strata are defined by the species, age, growth form, and size of the plants being considered. The first and most obvious stratum is made up of the trees present, the second is the shrub stratum, and the third is the herbaceous stratum. However, the tree stratum can be divided into a canopy (or large tree stratum) and a subcanopy (or small tree stratum). Moreover, there is another less obvious stratum in addition to the tree, shrub, and herbaceous strata.

5.2.1 From Seed to Tree

When a seed from a forest tree germinates, it gives rise to a germinal seedling that has the potential of ultimately growing into a tree large enough to occupy a position in the canopy stratum. However, the vast majority of germinal seedings do not survive, so this usually doesn't happen. In some forest communities, after seed fall and subsequent germination, germinal seedlings of some species of maple can be present at densities of several hundred per square meter. Chances of survival for a particular seeding are exceedingly low. Seedlings that do survive can transition through growth into saplings, and their chances of survival increase. Over time, it is possible for a sapling to grow into a tree, but mortality can occur at any stage of development from germinal seeding to canopy tree as a result of disease, herbivory, environmental stress, and both intraspecific and interspecific competition.

Fires represents another factor of some importance, especially in forests in which oaks are an important component. When a surface fire occurs at a particular site, either naturally as a result of lightning or by design (i.e., a prescribed or controlled burn), it removes much or all of the litter layer and reduces the cover of low shrubs. This makes the site more favorable for the establishment and growth of oaks.

The microsite where a seed from a tree falls to the ground can greatly influence its chances of survival. This is illustrated by what are known as "nurse logs." The latter are the decaying logs of fallen trees. These decaying logs are colonized by bryophytes (mostly mosses but also including some liverworts), which form an almost compete cover over the upper portion of the log. The presence of this cover of bryophytes promotes decomposition of the log but more importantly provides a favorable microsite for seed germination and the early development of the seeding. This is because the bryophytes provide a moist environment and a measure of protection. Nurse logs are not uncommon in the spruce forests found at higher elevations in the Appalachians.

5.2.2 Small Trees

Some species of trees found in Appalachian and Ozark forests do not possess the genetic potential to grow large enough to occupy the canopy and thus are commonly restricted to the subcanopy. Common examples that occur in the Appalachian and/or Ozarks are flowering dogwood (*Cornus florida*), sassafras (*Sassafras albidum*), American holly (*Ilex opaca*), striped maple (*Acer pensylvanica*), eastern red bud (*Cercis canadensis*), and service berry (*Amelanchier arborea*). These subcanopy trees coexist in the same three-dimensional ecological space and at the same time as trees that ultimately will grow large enough to reach the canopy. Although the two regions share most species of small trees in common, a few species such as the Ozark chinquapin (*Castanea ozarkensis*) occur only in the Ozarks. Certain other species such as mountain holly (*Ilex montana*) and striped maple are found only in the Appalachians (Stephenson et al. 1993; Stephenson 2013).

5.2.3 Shade Tolerance

Trees, as is true for all green plants, require sunlight in order to develop and grow. However, species differ with respect to just how much sunlight they require. The term shade tolerance refers to the capacity of a particular species to compete for survival, from seeding to mature tree, under the shaded conditions that exist beneath a forest canopy. There is considerable variation among species, and a few species cannot tolerate any shade (Valladares and Niinemets 2008). Trees that are moderately intolerant to shade include yellow poplar (*Liriodendron tulipifera*), black locust (*Robinia pseudoacacia*), and black cherry (*Prunus serotina*), while trees very tolerant to shade include Canadian hemlock (*Tsuga canadensis*), American beech (*Fagus americana*), and sugar maple (*Acer saccharum*). A larger group of species falls somewhere between these two extremes, and these trees display intermediate intolerance. Some examples are red maple (*Acer rubrum*), white ash (*Fraxinus americana*), white oak (*Quercus alba*), and northern red oak (*Quercus rubra*). These differences in shade tolerances are reflected in the structure and composition of forest communities.

Trees that are intolerant to shape are ill-equipped to become established beneath a forest canopy. Instead, they are early successional species that are adapted to colonize open areas where the canopy has been disrupted, such as in treefall gaps or after major disturbance (e.g., logging of the forest or treefall resulting from a high wind event) has taken place. These species have small airborne seeds or fruits, rapid growth rates, and the capability to become established in a wide range of different habitats.

5.2.4 Shrubs

Small trees and shrubs are not always easily distinguished from one another, and the shrub stratum is often defined largely on the basis of taxonomy. In other words, certain species are considered as shrubs largely by convention. Evidence for this is that a particular species may be listed as a shrub in some published regional floras and as a small tree in others (Radford et al. 1968; Strausbaugh and Core 1978; Gentry et al. 2013). Both trees and shrubs are woody plants, and the primary distinguishing feature relates to growth form. By definition, shrubs have multiple stems arising from the ground, whereas trees are characterized by a single primary stem (or trunk). Moreover, trees are recognized as being a larger size. However, this isn't always the case. It is not unusual for some species of shrubs (e.g., American witch hazel [*Hamamelis virginiana*]) to have a single stem, and certain other species (e.g., great laurel [*Rhododendron maximum*] in the Appalachians) to grow to a considerable size.

5.2.5 Vines

Vines (mostly woody but also including semi-woody or herbaceous examples) are defined as plants in which the stem requires support and thus is not free-standing. Some vines such as grapevines (*Vitis* spp.) and poison ivy (*Toxicodendron radicans*) climb the stems or trunks of other plants by producing twining stems, tendrils, or aerial roots. Other vines, including such examples as Virginia creeper (*Parthenocissus quinquefolia*), can climb a trunk or creep along the ground. Vines are not particularly common in relatively undisturbed deciduous forests but sometimes can be abundant in disturbed examples such as those in early successional regrowth after logging.

5.2.6 Herbaceous Plants

Plants recognized as members of the herbaceous stratum do not have woody tissues, although some small semi-woody plants (e.g., striped prince's pine [*Chimaphila maculata*] and teaberry [*Gautheria procumbens*]) occupy the same ecological space (Stephenson 2013). Most herbaceous plants are flowering plants, but lower vascular plants (especially ferns but also including lycopods in some situations) can be common and conspicuous elements of the herbaceous stratum. Many of the herbaceous flowering plants found in Appalachian and Ozark forests produce their flowers in spring, usually before the leaves are fully developed on the trees making up the canopy above them. These plants, often referred to as "spring wildflowers" or "spring ephemerals," are most apparent in moist sites, where their flowers can form an almost complete cover over the forest floor (Fig. 5.1).

Fig. 5.1 Spring ephemerals often form an almost complete cover over the forest floor in some mesic forests in the Appalachians. (Photo by Gary P. Fleming)

Zollner et al. (2005) listed 36 taxa of vascular plants that were endemic to the Ozarks. Although their definition of the Ozarks encompassed a slightly larger area than the one considered in this book, these are all plants that do not occur in the Appalachians. Most are perennial forbs (broadleaf herbaceous plants), a disproportionate number are restricted to or have distributions centered in the Ouachita Mountains, and an appreciable number are associated with open glades and not forests.

5.2.7 The Ground Layer

There is yet another stratum in the forests of the Appalachians and Ozarks (albeit less evident in the latter region) that is often overlooked or at least not considered. This stratum is best referred to as the ground layer. It consists of bryophytes (mostly mosses but including some liverworts) and lichens found on the surface of the soil, on rocks, coarse woody debris, and the trunks and branches of trees. In deciduous forests, bryophytes and lichens are largely restricted to substrates that are at least slightly elevated above the forest floor. These include the bases of living trees and the tops of fallen logs. The primary reason for this is the effect of the blanket of dead leaves that accumulates on the forest floor as a result of leaf fall.

5.3 Important Forest Trees

5.3.1 Oaks

Oaks (*Quercus*) and hickories (*Carya*) are consistently present in the forests of both the Appalachians and Ozarks (Braun 1950; Bryant et al. 1993; Stephenson et al. 2007). The species involved are often but not always the same, and even for species shared in common, their relative importance can be quite different in the two regions. Oaks found in both regions include white oak (*Q. alba*), which tends to be an important component of forests in both the Appalachians and Ozarks, northern red oak (*Q. rubra*), and black oak (*Q. velutina*). However, chestnut oak (*Q. montana*), one of the leading dominants in forests of the Appalachians, is not present in the Ozarks. In contrast, certain other species, such as chinquapin oak (*Q. muehenbergii*), blackjack oak (*Q. marilandica*), and post oak (*Q. stellata*) are relatively common in the Ozarks but rare in the Appalachians.

5.3.2 Hickories

Four species of hickory—pignut (*Carya glabra*), bitternut (*C. cordiformis*), mockernut (*C. alba*), and shagbark (*C. ovata*)—are found in both the Appalachians and Ozarks, but Texas hickory (*C. texana*) occurs only in the latter region. In general, hickories are more important in forests of the Ozarks than in those of the Appalachians, which reflects the fact that the Ozarks were mapped by Braun as falling within the Oak-Hickory Association (Braun 1950). Interestingly, one major distinction that the latter made in her famous monograph is that the trees of the western part of the Oak-Hickory Association, and thus the Ozarks, generally have a distinctly lower and shorter stature compared to those making up similar forests in the Appalachians.

5.3.3 Other Species of Trees

Among the other common and widespread trees found in both the Appalachians and the Ozarks are red maple (*Acer rubrum*), American beech (*Fagus grandifolia*), sugar maple (*A. saccharum*), cucumber tree (*Magnolia acuminata*), black locust (*Robinia pseudoacacia*), basswood (*Tilia americana*), and white ash (*Fraxinus americana*). However, the two regions differ in the presence or relative abundance of a number of other species. For example, tulip poplar (*Liriodendron tulipifera*), often the most abundant tree in the disturbed forests of lower slopes and valley floors in the Appalachians, does not occur naturally in the Ozarks. This also is the case for American chestnut (*Castanea dentata*), one of the dominant species in the forests of the Appalachians prior to the arrival of the chestnut blight fungus. Other trees found only in the Appalachians include striped maple (*Acer pensylvanicum*), Carolina hemlock (*Tsuga caroliniana*), black birch (*Betula lenta*), yellow birch (*B. alleghaniensis*), and big-leaved holly (*Ilex montana*) along with the small shrubby bear oak (*Quercus ilicifolia*). The overall diversity of trees in the Ozarks is lower than in the Appalachians, and species that occur only in the former region include maple-leaf oak (*Quercus acerifolia*), Ashe's juniper (*Juniperus ashei*), and Mexican plum (*Prunus mexicana*).

Broadleaf species of trees make up the greater portion of the forest cover in both the Appalachians and Ozarks, but some forests have an admixture of conifers and in others conifers are the dominant species present. For example, eastern white pine (*Pinus strobus*) is often a major component of forests in some portions of North Carolina, Tennessee, and Virginia. Throughout the Appalachians, the highest peaks and ridges are occupied by red spruce (*Picea rubens*). Only red spruce is present at most localities, but in a few examples either balsam fir (*Abies balsamea*) in the north or Fraser fir (*Abies fraseri*) in the south occurs along with red spruce (White et al. 1993). Canadian hemlock (*Tsuga canadensis*) is found along with red spruce at some localities as well as a few broadleaf species (e.g., yellow birch and American mountain ash [*Sorbus americana*]).

In the Ozarks, short-leaf pine (*Pinus echinata*) is the major conifer present, although eastern red cedar (*Juniperus virginiana*) occurs in some lowland, typically disturbed sites. In the Appalachians, short-leaf pine and red cedar are sometimes locally abundant but much less so than in the Ozarks. Two species of pine can be found in upland sites in the Appalachians that are absent from the Ozarks (Skeen et al. 1993). These are pitch pine (*P. rigida*) and Table Mountain pine (*P. pungens*). Forests in which either is the dominant species are never extensive and such forests are generally regarded as successional unless they occur on rock outcrops or in areas of very thin rocky soils. It is difficult for broadleaf trees to be become established on such sites.

5.4 Regional Differences

5.4.1 Seed Plants

Perhaps the most obvious floristic difference that exists between forests of the Appalachians and Ozarks is the composition of the shrub stratum. Species of blueberry (*Vaccinium*) and azalea (deciduous members of the genus *Rhododendron*) are common to abundant in the Appalachians but have a more limited distribution in the Ozarks (Fig. 5.2). Low-bush blueberry (*V. pallidum*) and deerberry (*V. stamineum*) are among the most widespread species in both regions. Farkleberry (*V. arboreum*) is sometimes present but not common in forests of the Ozarks, while species of azalea such as flame azalea (*R. calendulaceum*) are often abundant in forests of the Appalachians, where their colorful display of flowers is impossible to overlook in late spring (Fig. 5.3). In general, members of the family Ericaceae (the heath family) are simply more widespread and diverse in the Appalachians, with common examples such as mountain laurel (*Kalmia latifolia*) and minnie-bush (*Rhododendron pilosum*) absent from the Ozarks.

Among the more common vines in both the Appalachians and Ozarks are species of grape (*Vitis*), with fox grape (*V. vulpina*) and muscadine (*V. rotundifolia*) occurring with comparable levels of abundance in both regions. Summer grape

Fig. 5.2 Blueberries (*Vaccinium* spp.) are common in the forests of much of the Appalachians but have a more limited distribution in the Ozarks. (Photo by Steven L. Stephenson)

Fig. 5.3 The colorful flowers of flame azalea (*Rhododendron calendulaceum*) are difficult to miss in late spring in the forests of the Appalachians. (Photo by Steven L. Stephenson)

(*V. aestivalis*) is more common than either of these species in the Appalachians. Virginia creeper (*Parthenocissus quinquefolia*) and poison ivy (*Taxicodendron radicans*) are the two other widespread vines in both regions. Semi-woody species of greenbrier (*Smilax*) are often found in drier forests in both regions (Fig. 5.4). Common greenbrier (*S. rotundifolia*) and sawbrier (*S. glauca*) are the most common and widespread species.

The assemblage of spring wild flowers found in the Ozarks is less diverse than that of the Appalachians, although an appreciable number of species occur in both regions. These include spring beauty (*Claytonia virginica*), false Solomon's seal (*Polygonatum biflorum*), Dutchman's breeches (*Dicentra cucullata*), wild ginger (*Asarum canadense*), large-flower bellwort (*Uvularia grandiflora*), and Jack-in-the-pulpit (*Arisaema triphyllum*). However, there are some floristic differences. The yellow trout lily in the Ozarks is a different species (*Erythronium rostratum*) than the one (*E. americanum*) in the Appalachians, and the white trout lily (*Erythronium albidum*) is common in the Ozarks but rare in the Appalachians. Violets are common members of the herbaceous stratum in both regions, but the species involved tend to be somewhat different. Some species such as the common blue violet (*Viola*

Fig. 5.4 The semi-woody vines of greenbrier (*Smilax* spp.) are not uncommon in dry, open forests in both the Appalachians and Ozarks. (Photo by Steven L. Stephenson)

Fig. 5.5 Eastern teaberry (*Gaultheria procumbens*) is an example of a subshrub found in the Appalachians but absent from the Ozarks. (Photo by Charles Garratt)

sororia) and bird's-foot violet (*V. pedata*) are not uncommon in both regions. Other species, including the Canada violet (*V. canadensis*), three-part violet (*V. triparta*), and round-leaf violet (*V. roundifolia*), are either uncommon or absent from the Ozarks (Gentry et al. 2013).

Other species found in the Appalachians but not in the Ozarks include the subshrub eastern teaberry (*Gaultheria procumbens*) (Fig. 5.5), Canada mayflower (*Maianthemum canadense*), wild sarsaparilla (*Aralia nudicaulis*), American wintergreen (*Pyrola americana*), and whorled wood aster (*Oclemena acuminata*). The reverse is also the case, with other species occurring in the Ozarks but not in the Appalachians. Examples include Newton's larkspur (*Delphinium newtonianum*), Ozark cornsalad (*Valerianella ozarkana*), Arkansas blue star (*Amsonia hubrichtii*), Ozark mountain goldenrod (*Solidago ouachitensis*), dwarf spiderwort (*Tradescanta longipes*), and Ozark spiderwort (*T. ozarkana*).

Members of the genus *Trillium* occur in both regions, although they are much more common in the Appalachians (Fig. 5.6). The species involved are different, with toadshade trillium (*T. sessile*) the only species that is fairly common in both

Fig. 5.6 Trillium (*Trillium* spp.) is one of the best-known spring ephemerals in both the Appalachians and Ozarks. (Photo by Charles Garratt)

regions. Ozark trillium (*T. ozarkansis*), purple trillium (*T. recurvation*), and green trillium (*T. viridescens*) are the species most likely to be encountered in the Ozarks, while the large-flowered white trillium (*T. grandiflorium*), red trillium (*T. erectum*), and painted trillium (*T. undulatum*) are their ecological counterparts in the Appalachians (Radford et al. 1968; Gentry et al. 2013).

The herbaceous stratum in the more open, drier oak-hickory forests of the Ozarks is characterized by a different assemblage of species, some of which are common in the adjacent prairie region to the west. Members of this assemblage are bird's foot violet, slender mountain-mint (*Pycnanthemum tenuifolium*), late purple aster (*Symphyotrichum patens*), whorled milkweed (*Asclepias verticillata*), flowering spurge (*Euphorbia corollata*), and southern blazing star (*Liatris squarrosa*).

Most herbaceous plants are no more than about half a meter tall, but a few species greatly exceed this. Three of the most common examples are black cohosh (*Actaea racemosa*), American spikenard (*Aralia racemosa*), and Turk's cap lily (*Lilium superbum*), all of which sometimes can reach a height of more than two meters, and thus are taller than many shrubs. All three are not uncommon in moist forests of the Appalachians but American spikenard and Turk's cap lily have a more limited distribution in the Ozarks.

5.4.2 *Ferns and Lycopods*

Ferns and lycopods are significantly more important in the forests of the Appalachians, especially those at higher elevations, than in those of the Ozarks. This is presumably the result of the drier conditions that exist throughout the Ozarks. In a typical Ozark oak-hickory forest, Christmas fern (*Polystichum acrostichoides*) and ebony spleenwort (*Asplenium platyneuron*) are likely to be the only species present.

Christmas fern (Fig. 5.7) is the most common fern in the Appalachians and Ozarks as well as in the rest of the eastern United States. The species is evergreen,

Fig. 5.7 Christmas fern (*Polystichum acrostichoides*) is probably the most common fern in both the Appalachians and Ozarks. (Photo by Steven L. Stephenson)

and the fronds often remain green until Christmas, which is the reason for its common name. The fronds have 20–35 pairs of pinnae, and each pinna has a small triangular thumb-like lobe at the base. Christmas fern typically occurs in moist shady places and is particularly common on the banks along streams. However, in the Ozarks it is sometimes found in some surprisingly dry forests.

The ebony spleenwort (Fig. 5.8) is smaller than the Christmas fern (fronds 20–40 cm long in the former vs. 30–50 cm in the latter) and is also evergreen. The name is derived from its smooth dark, reddish brown petiole and rachis (midrib of the frond) of the narrow pinnate frond. There are two types of fronds: the fertile fronds with sori die in the winter, and the sterile fronds that remain evergreen and lie flat on the ground. This species is characteristic of dry rocky forests, although it also can occur in more moist habitats.

Ferns are a much more conspicuous element of the herbaceous stratum in the forests of the Appalachians, where such species as the evergreen shield fern (*Drypteris intermedia*) and hay-scented fern (*Dennstaedtia punctilobula*) commonly occur. No species of lycopod is particularly common in the forests of the Ozarks, but lycopods can be so abundant that they sometimes form an almost complete cover over the forest floor at higher elevations in the Appalachians (Fig. 5.9). The most common species are common clubmoss (*Lycopodium clavatum*), stiff clubmoss (*Spinulum annotinum*), and tree clubmoss (*Dendrolycopodium obscurum*).

5.5 Nonvascular Plants

The nonvascular plants—bryophytes and lichens—that make up the ground layer rarely reach a height of more than a few centimeters but in high-elevation spruce and spruce-fir forests in the Appalachians bryophytes sometimes cover extensive areas of the forest floor (Carvel 1993). They are less conspicuous in forests at lower

Fig. 5.8 The ebony spleenwort (*Asplenium platyneuron*) is common in forests of both the Appalachians and Ozarks. (Photo by Kim Fleming)

Fig. 5.9 Clubmosses (species in several genera including *Lycopodium*) are often abundant in high-elevation coniferous forests of the Appalachians. (Photo by Jason Hollinger)

elevations in the Appalachians and even more so in the forests of the Ozarks, presumably because of the drier conditions that prevail throughout the latter region. In deciduous forests, this stratum is not readily apparent during the summer, but it becomes far more noticeable during winter when most of the higher plants present in a forest have lost their leaves.

5.5.1 Mosses and Liverworts

In the habitats where they occur, species of bog moss (*Sphagnum* spp.) are usually the dominant plants present (Fig. 5.10). Members of the genus, also called peat mosses, are typically associated with bogs and mires, with the former common at higher elevations in the Appalachians. However, patches of bog moss can occur in moist microsites on the forest floor. There are approximately 380 species of bog moss worldwide, most of which are difficult to distinguish. This is especially true when several different species occur together in the same bog, which is usually the case. The plant body of a bog moss has a unique cellular structure that allows it to absorb water, and large masses of the moss can retain huge quantities of water. Quantitatively, these plants can hold many times as much as their dry weight. Accumulations of bog moss at a particular site provide a special habitat for numerous plants that could not survive elsewhere. Bog moss acidifies its surroundings, and the plant does not readily decay because of this. As a result, dead bog moss plants can accumulate to depths of several meters, forming what is called peat. When collected and dried, peat can be burned as fuel. Since a *Sphagnum*-dominated bog is capable of spreading, it can encroach upon the surrounding forest.

Leafy liverworts are among the most overlooked plants in forests, although small examples are often common on moist, decaying logs and the bark of trees. A decorticated (having lost its bark) red spruce, fir or hemlock log in high-elevation forests in the Appalachians often can have large areas of its surface covered with leafy liverworts. The three-lobed bazzania (*Bazzania trilobata)* is a large leafy liverwort that

Fig. 5.10 Patches of peat moss (*Sphagnum* spp.) often occur on moist areas of the forest floor in high elevation forests in the Appalachians. (Photo by Steven L. Stephenson)

Fig. 5.11 Mats of the three-lobed bazzania (*Bazzania trilobata*) are typically found at the bases of trees in high-elevation red spruce forests in the Appalachians. (Photo by Steven L. Stephenson)

is common in high-elevation forests of the Appalachians, where it occurs in large bright green mats around the bases of trees and over areas of the forest floor (Fig. 5.11). The species also is common in moist shaded ravines. The three-lobed bazzania is rare in the Ozarks, where the few records are from scattered localities in the northern portion of the region. The leaves on this plant are recurved and thus point downward. Each leaf has three short teeth, which accounts for both the common name and specific epithet. The three-lobed liverwort sometimes has been called the "millipede weed" for the resemblance of the plant to this segmented arthropod.

Among the more common mosses in the forests of both the Appalachians and Ozarks are the bordered mnium (*Mnium marginatum*), pincushion moss (*Leucobryum albidum*), *Calliergonella curvifolia*, haircap moss (*Polytrichum* spp.), delicate fern moss (*Thuidium delicatulum*), and windblown moss (*Dicranum scoparium*). Certain other species, particularly those associated with high-elevation coniferous forests in the Appalachians, do not occur in the Ozarks. For example, this is the case for the mountain fern moss (*Hylocomium splendens*), which is characteristic of more northern areas of the United States and Eurasia. A similar distribution pattern, but limited solely to eastern North America, is *Racomitrium aduncoides*, ranging from Maine down to high elevations in North Carolina and Tennessee. There are other species that are restricted solely to higher elevations in the Appalachians, such as *Leptohymenium sharpii*, and others that are disjunct from higher elevations in Central America, such as *Donrichardsia pringlei*. Most of these highly restricted mosses are sufficiently rare that they don't have common names.

On the other hand, there are mosses in the Ozarks that are more adapted to drier habitats and do not extend to the Appalachians (Redfearn 1972). For example, *Venturiella sinensis* var. *angusti-annulata* is known only from the western Ozarks. Similarly, a number of species, adapted to drier habitats, extend from the western United States as far east as the Ozarks but not to the Appalachians. An example is *Ptychomitrium sinense* (previously known as *P. leibergii*), known in North America only from the western Ozarks and adjacent western North America.

Fig. 5.12 Haircap mosses (*Polytrichum* spp.) occur in several different types of forested and non-forested habitats in both the Appalachians and Ozarks. (Photo by Steven L. Stephenson)

Mosses are usually small plants that grow close to the ground. However, the haircap mosses (*Polytrichum* spp.) represent somewhat of an exception, since they can reach a height of 20 cm when the sporophyte is present (Fig. 5.12). These mosses get their name from the hairs that cover the hood-like cap (calyptra) that encloses the spore-containing capsule during development. The capsule and the stalk (seta) that holds it aloft are two parts of the sporophyte generation in the life cycle. The other generation (the gametophyte) is the green leafy plant located at the base of the seta. Vascular plants also have these two generations in their life cycles, but only the sporophyte generation is visible. Although the haircap moss, like all mosses, is nonvascular, it does possess a type of tissue in the main axis of the body ("stem") that is capable of conducting water from the base to aerial portions of the plant. This tissue is analogous to the xylem of vascular plants. *Polytrichum* is a genus of about 70 species, and they occur worldwide. Haircap moss occurs in more open areas than most mosses and is typically found on elevated microsites, including the top of a stump or in bogs on top of a hummock.

As a general observation, bryophytes are more abundant in the relatively moister forests of the Appalachians than in the Ozarks, where these organisms tend to be confined largely to sites (e.g., shaded ravines) where unusually moist conditions exist. However, careful examination of the bases of trees, the tops of decaying logs on the forest floor, and crevices among rocks will yield some representatives of this group of organisms. The greater geological diversity of the Ozarks provides more kinds of rock habitats than do the Appalachians.

5.5.2 Lichens

Lichens are widespread and locally often abundant in forests (Dey 1993), including those of the Appalachians and Ozarks. Although this fact is not immediately obvious when a lichen is observed in the field, lichens are "composite" organisms and consist of a fungus and one or more species of algae (or cyanobacteria) living together in an intimate biological association. The structure recognized as a lichen bears little resemblance to either organism and can survive under environmental conditions that would be fatal to either of the organisms of which it is composed. Because the algal component (or photobiont) requires light for photosynthesis, lichens cannot survive where light is limited, such as beneath a dense canopy of a coniferous forest.

Lichens are classified into three somewhat overlapping growth forms—crustose (flat and crust-like), foliose (leaf-like), and fruticose (having a three- dimensional or "shrubby" appearance). All three growth forms can occur on the types of substrates on which lichens are found. These are exposed areas of bare soil, pieces of coarse woody debris, the bark surfaces of living trees, and rock outcrops. In forests where there are large gaps in the canopy, lichens occur along with bryophytes on the forest floor. In these situations, fruticose lichens such as reindeer moss (*Cladina* spp.) are not uncommon. Foliose lichens such as shield lichens (*Parmelia* spp.) are often conspicuously present on the bark of living trees in most forests and also occur along with various crustose lichens on rock outcrops. Perhaps the most distinctive of the foliose lichens found on rock outcrops are species of rock tripe (*Umbilicaria* spp.). Since most lichens are widely distributed, the assemblages of species found in the Appalachians and Ozarks would be expected to be fairly similar, although detailed taxonomic comparisons appear to be lacking.

One of the more conspicuous lichens in both the Appalachians and Ozarks is the British soldier lichen (*Cladonia cristatella*). The British soldier lichen is a fruticose lichen and received its common name from the regimental coats that the British soldiers wore during the American Revolution (Fig. 5.13). The vegetative portion of the lichen (or thallus) consists of a patch of small, green scalelike structures. These give rise to the upright reproductive structures, which often branch at the tip. Each branch has a rounded, ball-like, spore-bearing structure at the top. This is bright red, the same color of the regimental coats of the British soldiers. The British soldier lichen is found on substrates (rocks, the soil, or decaying logs and stumps) that are exposed to sunlight, which the algal component of the lichen needs in order to photosynthesize. Lichens like this one play a subtle yet important role in nature. Their presence on a particular substrate enhances the colonization of the substrate by other organisms. As such, they promote succession.

Fig. 5.13 The British soldier lichen (*Cladinia cristatella*) is one of the more conspicuous lichens found in both the Appalachians and Ozarks. (Photo by Kasia Halka)

5.5.3 *Fungi*

Several days after a heavy rain, any visit to a forest in the Appalachians or Ozarks during the period of late spring to early fall will reveal the presence of the fruiting bodies of another group of organisms—the fungi. Some people visit forests for the purpose of collecting the fruiting bodies (often referred to as mushrooms) of certain species that are known to be edible. However, the fungi are vastly more important than their use as a type of "wild" food. This and other aspects of fungi will be discussed in Chap. 9.

References

Braun EL (1950) Deciduous forests of eastern North America. Blakiston Company, Philadelphia

Bryant WS, McComb WC, Fralish JS (1993) Oak-hickory forests (Western Mesophytic/Oak-Hickory forests). In: Martin WH, Boyce SC, Echternacht AC (eds) Biodiversity of the Southeastern United States: upland terrestrial communities. John Wiley and Sons, New York, pp 143–201

Carvel KL (1993) Bryophytes and their ecological role in the upland forests. In: Stephenson SL (ed) Upland forests of West Virginia. McClain Printing Company, Parsons, WV, pp 87–100

Dey JP (1993) Observations on the fruticose and foliose lichens of the upland forest region. In: Stephenson SL (ed) Upland forests of West Virginia. McClain Printing Company, Parsons, WV, pp 101–124

Gentry JL, Johnson GP, Baker BT, Witsell CT, Ogle JD (2013) Atlas of the vascular plants of Arkansas. PMC Solutions, Fayetteville, AR

Radford AE, Ahles HE, Bell CR (1968) Manual of the vascular Flora of the Carolinas. University of North Carolina Press, Chapel Hill

Redfearn PL (1972) Mosses of the interior highlands of North America. Ann Mo Bot Gard 59:1–104

Skeen JN, Doerr PD, van Lear DH (1993) Oak-hickory-pine forests. In: Martin WH, Boyce SC, Echternacht AC (eds) Biodiversity of the Southeastern United States: upland terrestrial communities. John Wiley and Sons, New York, pp 1–33

Stephenson SL (2013) A natural history of the Central Appalachians. West Virginia University Press, Morgantown, WV

Stephenson SL, Ash AN, Stauffer DF (1993) Appalachian oak forests. In: Martin WH, Boyce SC, Echternacht AC (eds) Biodiversity of the Southeastern United States: upland terrestrial communities. John Wiley and Sons, New York, pp 255–303

Stephenson SL, Adams HS, Huebner CD (2007) Upland forest vegetation of the Ozark Mountains in northwestern Arkansas. Rhodora 109:197–221

Strausbaugh PD, Core EL (1978) Flora of West Virginia. Seneca Books, Morgantown, WV

Valladares F, Niinemets Ü (2008) Shade tolerance, a key plant feature of complex nature and consequences. Annu Rev Ecol Evol Syst 39:237–257

White PS, Buckner ER, Pittillo JD, Cogbill CV (1993) High-elevation forests: northern hardwoods forests and associated communities. In: Martin WH, Boyce SC, Echternacht AC (eds) Biodiversity of the Southeastern United States: upland terrestrial communities. John Wiley and Sons, New York, pp 305–337

Zollner D, MacRoberts MH, MacRoberts BR, Ladd D (2005) Endemic vascular plants of the interior highlands. Sida 21:1781–1791

Chapter 6
Non-forested Habitats of the Appalachians and Ozarks

Abstract Prior to the arrival of the Europeans, a large portion of the Appalachians and Ozarks was covered in forests. However, the forest cover was not complete because there were naturally occurring open areas, especially in the Ozarks. These open areas included examples such as grassy glades, rocky glades, open streamside habitats, fens, cliff faces, and rock outcrops that were common to both regions along with others (e.g., shale barrens and bogs) either generally characteristic of or restricted to the Appalachians. There also was a range of habitats such as woodlands and savannas that included trees but were not closed canopy forests. The purpose of this chapter is to describe each of these types of non-forest habitats, including the plants associated with them.

Keywords Non-forested habitats · Glades · Balds · Bogs · Rock outcrops · Barrens · Rare plants · Cliff faces

6.1 Introduction

Prior to the arrival of the Europeans, a large portion of the two regions considered in this book was covered in forests (Davis 1983). The forest cover was not complete in the Ozarks because there were areas of naturally occurring grasslands, open streamside habitats, and rock outcrops. In the Appalachians, forest cover was interrupted by several different types of open areas. Rock outcrops were not uncommon on the summits and upper slopes of mountains and also occurred on the lower slopes of deep stream valleys and gorges. There also were small areas scattered throughout the region occupied by special habitats such as glades, balds, shale barrens, cedar glades, and bogs. All of these were largely or completely devoid of trees. The purpose of this chapter is to consider each of these non-forested habitats as they currently exist.

S. L. Stephenson, R.-M. Muzika, *Forests of the Southern Appalachian Highlands and Interior Highlands of the United States*, Plant and Vegetation 20,
https://doi.org/10.1007/978-3-032-01155-8_6

6.2 Glades, Balds, and Barrens

The term "glade" has been used for several different types of habitats. In West Virginia, locals applied the term to bogs such as Cranberry Glades. However, it is more appropriate for certain other types of habitats, the more common of which are grassy glades, gravel (or rock-gravel) glades, and cedar glades (Bartgis 1985, 1993). However, a glade in the Ozarks generally refers to an open area, typically on an exposed aspect (i.e., a south or west facing slope where bedrock is close to the surface). The characteristics they all share in common are no or very low tree cover, thin rocky soils, xeric conditions, high soil temperatures during summer, and the predominance of herbaceous plants, especially warm-season (i.e., C4) grasses.

6.2.1 Grassy Glades

Grassy glades are found in both the Appalachians and Ozarks but are much more common in the latter region, where the assemblage of plants present more closely resembles that of the western prairies than the glades of the Appalachians (Fig. 6.1). The size of glades ranges from many hectares to less than 100 m^2, but most are relatively small, ranging from one to three hectares. The usual situation is for a glade to be surrounded by forest where the latter has not been disturbed or removed. A transition zone, or ecotone, where the forest becomes increasingly open as it transitions into a glade is characteristically present. In the Ozarks, grassy glades are often transitional habitats that intergrade into tall grass prairies.

Fig. 6.1 Grassy glade with conspicuous wildflowers (*Echinacea purpurea*) in the Caney Mountain Conservation Area of southwestern Missouri. (Photo by Susan Farrington)

6.2.2 Open Glades

Open glades in the Ozarks can occur on sandstone, igneous rocks, chert, dolomite or limestone and the type of rock largely determines the composition of the plant community present. Sandstone, chert, and igneous rocks have appreciably less alkaline soils than those occurring over limestone or dolomite, and thus support species associated with acidic soils. In contrast, limestone and dolomitic glade support species associated with alkaline or neutral soils. Regardless of the type of rock, open glades are characterized by thin soils (Ware 2002). In some respects, these glades form a category of their own, since grasses are not the most prominent plants (Jeffries 1985, 1987). Instead, the usual dominants are such forbs as big-flowered coreopsis (*Coreopsis grandiflora*) and one-seed croton (*Croton willdenowii*). Grasses that are present include poverty grass (*Aristida dichotoma*) and switch grass (*Panicum virgatum*).

Grassy glades in the Ozarks are less xeric than rock-gravel glades, have deeper soils than rock-gravel grades, and support a more diverse assemblage of plants (Gentry et al. 2013). Characteristic plants include naked-stem sunflower (*Helianthus occidentalis*), prairie sunflower (*Helianthus pauciflorus*), and prairie violet (*Viola pedatifidia*). However, the most important plants present are various warm-season grasses, including broomsedge bluestem (*Andropogon virginicus*), big bluestem (*A. gerardii*), and poverty oak grass (*Danthonia spicata*). A few of the species found in these glades are endemic to the Ozarks, including Bush's purple coneflower (*Echinacea paradoxa*), Bush's skullcap (*Scutellaria bushii*), and Trelease's larkspur (*Delphinium treleasei*). The species of plants associated with grassy glades were described by Baskin and Baskin (2000).

Dolomitic glades are the most common type of glade in the Ozarks, but igneous glades also are conspicuously present as well. Missouri bladderpod (*Lesquerella filiformis*) and celestial lily (*Nemastylis nutalli*) are among the endemic species occurring on dolomitic glades, and Mead's milkweed (*Asclepias meadii*), a federally listed species, occurs on igneous glades. Animals associated with glades in the Ozarks include the roadrunner (*Geococcyx californianus*), eastern collared lizard (*Crotaphytus collaris*), the Texas brown tarantula (*Aphonopelma hentzi*), and several species of insects. Forests surrounding igneous glades are likely to consist of shortleaf pine (*Pinus echinata*), while dolomitic or limestone glades are mostly surrounded by or adjacent to forests containing chinquapin oak (*Quercus muehlenbergii*) or gum bumelia (*Bumelia lanuginosa*.)

More than 160,000 ha (400,000 acres) of glade habitat are estimated to occur in just the Missouri Ozarks, and much of it has been degraded (i.e., forests have encroached upon the glade habitat or changes in land use have occurred). Natural disturbance regimes in the Ozarks include those associated with moderately frequent, low intensity fire, which in conjunction with the shallow soil and xeric conditions, minimize colonization by plant species other than glade specialists. The rapid expansion of the eastern red cedar (*Juniperus virginiana*) on glade habitats represents the most persistent ecological concern and requires intervention and removal to reclaim the glade habitat. Ecological restoration of glade habitats that include removal of the target species and the application of fire has steadily expanded glade habitats and protected glade species.

6.2.3 Balds

Balds have a limited distribution in the Appalachians (Rentch and Fortney 1997) and true balds are not considered to occur in the Ozarks. As defined herein, a bald is habitat restricted to areas on high elevation (typically >1200 m) summits and upper slopes of mountains where trees are conspicuously absent (Fig. 6.2) These habitats are generally characterized by thin, rocky soils and occur in exposed, windswept situations (Mark 1958). In its more common expression, the vegetation present is dominated by shrubs, including Catawba rosebay, mountain laurel (*Kalmia latifolia*), minnie-bush (*Rhodendron pilosum*), and several species of blueberry (*Vaccinium*). Such balds are referred to as shrub balds as opposed to grass (or grassy) balds, in which grasses (most notably Mountain oak grass [*Danthonia compressa*]) and various other herbaceous plants are the predominant elements in the vegetation. There are two primary features that distinguish grassy balds from grassy glades. The first is topographic position (i.e., high-elevation areas at or near the summit of a mountain for grassy balds versus lowland or intermediate elevations for grassy glades). The second is the underlying rock, with grassy balds occurring over sandstone or various metamorphic rocks and grassy glades mostly occurring over limestone or dolostone. As a result, soil pH is appreciably higher for grassy glades.

Today, shrub balds are much more common than grassy balds, in part because the latter are more prone to succession. For example, the well-known grassy balds in the Great Smoky Mountains are being rapidly encroached upon by trees. The origin and ecological dynamics of balds have received considerable study but remain poorly understood (Weigl and Knowles 1995). On the most environmentally severe sites, where soils are very thin, with often large boulders present, succession is likely to

Fig. 6.2 Grass bald on Whitetop Mountain in the southern Blue Ridge of the Appalachians. (Photo by Gary P. Fleming)

proceed very slowly. As such, some of the shrub balds found on such sites are likely to have persisted for a very long time. Others owe their existence to the wholesale logging and subsequent fires that occurred in the late 1800s and early 1900s. It is generally agreed that disturbance factors, including fire and grazing by livestock, almost certainly have and will continue to play an important role in the persistence of balds.

As noted elsewhere in this book, the high-elevation forests of the Appalachians have been subjected to extensive logging and often subsequent fires (Clarkson 1964). The post-logging succession that took place in areas once dominated by red spruce and Canadian hemlock often produced what are known as heath balds. These are typically dominated by a dense growth of great laurel on more moist sites and by mountain laurel on more xeric sites. Scattered trees are usually present in most situations. Heath balds (especially those involving mountain laurel) are often so dense as to be virtually impassable to humans.

In the Ozarks, the term "bald" is used colloquially to refer to an area in southwestern Missouri and northwestern Arkansas where there is a series of semicontinuous dolomitic glades that appear to occupy an entire ridge and upper slopes at a particular locality. This region, located on the southern flank of the Ozark uplift, includes numerous extended ridges and rounded knobs.

6.2.4 Shale Barrens

Shale barrens represent another distinctive habitat found in the northern portion of the Appalachians as the region is delimited this book and one that is classified as a globally rare (Platt 1951). Shale barrens typically occur at intermediate to low elevations (300–850 m) on steep slopes with a southern exposure and underlain by highly friable shale. The typical situation is for a shale barren to occur on steep slope with a stream at the base (Fig. 6.3). The highly friable shale produces an unstable substrate for the development of soil and the establishment of plants.

Fig. 6.3 Shale barren in the Ridge and Valley of the Appalachians. (Photo by Irvine T. Wilson)

Fig. 6.4 Millboro leatherflower (*Clematis viticaulis*) is a shale barren endemic found in the Ridge and Valley of the Appalachians. (Photo by Steven L. Stephenson)

Fig. 6.5 Kates-mountain clover (*Trifolium virginicum*) is another shale barren endemic found in the Ridge and Valley of the Appalachians. (Photo by Charles Garratt)

Because of this, the slope receives direct sunlight, creating surface temperatures during summer that are lethal for most plants. However, some species of plants have evolved to become adapted to the unusually harsh conditions. Among these are the shale-barren evening primrose (*Oenothera argillicola*), shale-barren ragwort (*Packera antennariifolia*), shale barren buckwheat (*Eriogonum allenii*), white-haired clematis (*Clematis albicoma*), shale barren rock cress (*Boechera serotina*), Millboro leatherflower (*Clematis viticaulis*) (Fig. 6.4), and Kate's-mountain clover (*Trifolium virginicum*) (Fig. 6.5).

The species listed above are considered to be shale barren endemics because they are restricted to this habitat, and as a result of this fact are regarded as rare or in the case of shale barren rock cress endangered. No other type of habitat in the Appalachians supports such a high proportion of endemics. The type locality for many of the species listed above is a shale barren on Kate's Mountain near White Sulphur Springs in West Virginia (Core 1966). Non-endemics found on shale barrens include bird's foot violet (*Viola pedata*), Pennsylvania sedge (*Carex pensylvanica*), moss phlox (*Phlox subulata*), mountain nailwort (*Paronychia montana*), dwarf cinquefoil (*Potentilla canadensis*), smooth sunflower (*Helianthus laevigatus*), and hairy wood mint (*Blephilia ciliata*).

Shale barrens also occur in the Ozarks but are much less well-known and have a more limited distribution. The best examples occur in the Ouachita Mountains of Arkansas. As is the case in the Appalachians, these shale barrens occur on steep south- and west-facing slopes where there are outcrops of shale. Plants that are largely or completely restricted to these habitats include Pelton's rose-gentian (*Sabatia arkansana*), laceflower (*Ptilimnium nuttallii*), lance-leaved ragweed (*Ambrosia bidentata*), Drummond's St. John's-wort (*Hypericum drummondii*), Ozark calamint (*Calamintha arkansana*), and threadleaf evening primrose (*Oenothera linifolia*).

Although shale barrens are considered as non-forested areas, a few trees can colonize less unstable areas, so it is not a completely open habitat. In the Appalachians, the trees most likely to be encountered are usually stunted examples of Virginia pine (*Pinus virginiana*), chestnut oak (*Quercus montana*), and pignut hickory (*Carya glabra*). However, red cedar (*Juniperus virginiana*) is consistently present on somewhat richer shales in the northern part of the Ridge and Valley portion of the Appalachians.

6.2.5 Cedar Glades

Cedar glades occur in both the Appalachians and Ozarks but are more common and better studied in that portion of the United States occurring between these two regions (Baskin and Baskin 2000, 2004). Two types of cedar glades have been recognized on the basis to topographic position. These are "flatrock" cedar glades that are characteristic of relatively flat terrain and "hillslope" cedar glades found on steep, south- to southwest-facing slopes. The latter situations are similar to those on which shale barrens occur. As suggested by their very name, cedar glades are distinguished by the presence of red cedar (*Juniperus virginiana*). Although present, red cedar typically occurs at low densities, so the overall habitat is very open. However, this tree often occurs in areas adjacent to the glade and has the potential of increasing expanding enough over time to convert the glade into a red cedar-dominated community. Some of the species associated with cedar grades in the Ozarks are prairie dock (*Silphium integrifolium*), Missouri primrose (*Oenothera speciosa*), and eastern prickly-pear (*Opuntia humifusa*). In the Appalachians, little bluestem (*Schizachyrium scoparium*), side oats grama grass (*Bouteloua curtipendula*), forest goldenrod (*Solidago arguta*), Blue Ridge ragwort (*Packera millefolium*), button snake-root (*Eryngium yuccifolium*), Canadian summer bluet (*Houstonia canadensis*), woodland sunflower (*Helianthus divaricatus*), and wild bergamot (*Monarda fistulosa*) are some of the more characteristic species.

The most immediate threat to all types of glade habitats, as is ultimately the case for all natural habitats, is the impact of humans. Although the sites on which glades occur are usually not suitable for agriculture, they are subject to be used in a number of the other ways such as residential, industrial, and infrastructure development (e.g., roads).

6.3 Bogs, Fens, and Riparian Zones

6.3.1 Bogs

Perhaps the most interesting non-forested habitats in the Appalachians are bogs. True bogs are not found in the Ozarks and occur only at scattered localities in the Appalachians. Bogs occupy areas on flat or gently sloping slopes where groundwater discharge, seepage from adjacent mountain slopes, or impeded stream flow results in unusually moist conditions (Wieder et al. 1981). Cold air drainage from adjacent slopes also helps to maintain cool conditions for the bog. Most bogs are characterized by the dominance of bryophytes, especially species of peat moss (*Sphagnum*) and haircap moss (*Polytrichum*). Although bogs are non-forested habitats, they are almost invariably surrounded by forests, either broadleaf trees in examples found at lower elevations or conifers (usually red spruce [*Picea rubens*]) for those located at the highest elevations. It is not uncommon for the vegetation of a bog to form a mosaic consisting of relatively flat open areas dominated by sphagnum mosses, scattered patches of ferns or other vascular plants, slightly elevated areas that support the growth of shrubs or a few small trees, an outer border lined with shrubs, and the very edge of a surrounding forest. Various sedges, including cotton grass (*Eriophorum virginicum*, which is actually a sedge despite its common name), are consistently present throughout most bogs. Common trees at the edge of a bog include red spruce and Canadian hemlock (*Tsuga canadensis*) in bogs at higher elevations, while trees associated the edges of lower elevation bogs are more likely to be such species as red maple (*Acer rubrum*) and northern red oak (*Quercus rubra*). Some of the most characteristic shrubs found along the edges of bogs are the great laurel (*Rhododendron maximum*), Catawba rosebay (*R. catawbiense*), smooth alder (*Alnus serrulata*), long-stalked holly (*Ilex collina*), and wild raisin (*Viburnum cassinoides*).

Most bogs are relatively small, usually less than 0.4 ha, but the very largest examples can exceed 10 ha (Francl et al. 2004; Schultheis et al. 2010). For example, Cranberry Glades in West Virginia extends over 300 ha (Fig. 6.6). Small patches of sphagnum mosses occur at numerous places in the high-elevation forests of the Appalachians where water collects and the soil is unusually moist. However, these are not considered to be true bogs. Some small bogs may occur directly over mineral soil, but the usual situation is for a layer of peat to develop beneath the sphagnum mosses. This layer of peat can be quite thick, sometimes exceeding a meter. Soils in bogs are extremely acidic and thus limit the occurrence of those plants that cannot survive in this type of habitat.

Bogs are "biological islands" since the assemblage of species present is markedly different from that of the surrounding forest (Pittillo 1994). Some of the species present in bogs are rare or disjuncts that occur far south of their usual range. Some noteworthy examples are round-leaf sundew (*Drosera rotundifolia*), bog willow-herb (*Epilobium leptophyllum*), three-seeded sedge (*Carex trisperma*), large cranberry (*Vaccinium macrocarpon*), small cranberry (*V. oxycoccus*),

Fig. 6.6 Cranberry Glades is a large well-known bog in the Allegheny Mountains of West Virginia. (Photo by Steven L. Stephenson)

yellow-fringed orchid (*Habenaria ciliaris*), cottongrass (*Eriophorum virginicum*), rose pogonia (*Pogonia ophioglossoides*), and purple-fringed orchid (*H. fimbriata*). Purple pitcher plant (*Sarracenia purpurea*) is found in a few bogs in the Central Appalachians, but this species is not native, having been introduced in the early 1900s. Ferns tend to be abundant both in and around bogs, with cinnamon fern (*Osmundastrum cinnmomeum*) especially conspicuous because of its large size.

The successional dynamics of bogs are not completely understood. Some examples are clearly becoming smaller as a result of the invasion of shrubs and trees. However, there are certain other factors (e.g., fire, beavers, and various types of human disturbance) that can counteract this pattern. The possible long-term effect of climate change on bogs has yet to be determined, but it is likely to be detrimental to many of the organisms that live in this habitat and indeed to the habitat itself. It seems likely that acid rain has already had some impact on high-elevation bogs.

Bogs now occur at some localities in the Appalachians where they did not exist prior to extensive logging. The classic and also the largest example is Canaan Valley in West Virginia (Fig. 6.7). Canaan Valley is an elongated, oval-shaped valley nestled high in the Allegheny Mountains in Tucker County, West Virginia (Fortney 1993). Legend has it that Canaan Valley was named by George Casey Harness, a Bible reading, bear hunting explorer, who made numerous expeditions into the region. Harness is said to have been so impressed with the beauty of the valley that the cried, "Behold! The land of Canaan! Canaan is pronounced locally as "ka-nané" and not the Biblical pronouncement of "Ká-nan."

The original vegetation of Canaan Valley was an almost coniferous forest dominated by red spruce and Canadian hemlock. An admixture of balsam fir occurred in some of the wetter places, and great laurel was consistently present in the understory. Non-forested areas occurred in only a few places. This forest suffered the fate as almost all forests throughout most of the Allegheny Mountains—it was completed logged during the late 1800s and early 1900s. The succession that following logging resulted in a dramatic change in the vegetation. Much of the valley floor

Fig. 6.7 Canaan Valley in the Allegheny Mountains of West Virginia (photo by Steven L. Stephenson)

became a mosaic of wetlands. The total area of these wetlands (approximately 24,000 ha) is greater than any other wetland in the entire Appalachian region. Bogs cover approximately 570 ha of the 24,000 ha. The coniferous forests that occurred on the slopes of the valley have been mostly replaced by northern hardwood forests, which also have developed on some the higher areas on the valley floor. The bogs contain many of the same species as other bogs in the region, with dewberry (*Rubus hispidus*), bog goldenrod (*Solidago uliginosa*), beaked sedge (*Carex rostrata*), and cottongrass (*Eriophorum virginicum*) the most common.

Canaan Valley was once targeted to be a pump storage reservoir by Monongahela Power Company, but this project came under severe pressure from environmental groups and ultimately didn't happen. Today, a portion of the valley is taken up by housing developments. However, large wetland areas in the valley are now afforded protection in the Canaan Valley National Wildlife Refuge and Canaan Valley Resort State Park.

6.3.2 Fens

Fens are another but much less common type of wetland habitat found at scattered localities in both the Appalachians and Ozarks. While bogs are underlain by sandstones, fens are underlain by limestone or dolostone. As a result, they represent a much less acidic environment for the plant present. Fens typically occur in poorly drained natural depressions, on gentle slopes, or as seepage areas along streams. They may be permanently wet to intermittently dry and are generally fed by seepage from calcium-rich groundwater. Like bogs, the vegetation usually occurs as a mosaic consisting of areas dominated by shrubs and various herbaceous plants, with trees around the edges. Some of the more characteristic plants of Appalachian fens are Canadian burnet (*Sanguisorba canadensis*), short-tailed rush (*Juncus subcaudatus*), slender spikerush (*Eleocharis tenuis*), grass-of-Parnassus (*Parnassia grandifolia*), and sensitive fern (*Onoclea sensibilis*).

Ozark fens are distinguished by the degree of soil moisture as well as the plants associated with a particular fen. Soils of Ozark fens are mucky and saturated, whereas prairie and forested fens arise from both ponding of rainwater and calcareous seepage. Ozark fens are highly variable in shape and size but generally occur on lower slopes with neural to somewhat alkaline soil (Nelson 2010). Restricted plants of Ozark fens include grass pink (*Calopogon tuberosus*) and rose pogonia (*Pogonia ophioglossoides*). Plants typically associated with Ozark fens include Riddell's goldenrod (*Solidago riddellii*), Virginia mountain-mint (*Pycnanthemum virginianum*), yellow-loosestrife (*Lysimachia quadriflora*), and swamp lousewort (*Pedicularis lanceolata*). The last three species also occur in Appalachian fens. Some of the plants found in fens, such as Canadian burnet in the Appalachians, are rare.

Forested fens, although representing mostly open areas, often occur within forested landscapes since these are associated with riparian areas such as small floodplains. These may include elevated hummocks that support distinct plant communities. Forested fens are considered a relatively rare community type but contain two restricted plants, leafy bulrush (*Scirpus polyphyllus*) and wild sweet William (*Phlox maculata* ssp. *pyramidalis*). These fens are more hydrologically dynamic than Ozark fens.

Unfortunately, fens have long been subject to disturbance, especially by human activities. Of particular importance are historic draining efforts and the encroachment of invasive plants. The Grasshopper Hollow Natural Area in the Mark Twain National Forest contains the largest known fen complex in the unglaciated portion of North America.

6.3.3 Inland Salt Marsh

One truly unique non-forested habitat—an inland salt marsh—is located near the town of Saltville, Virginia in the Appalachians. The marsh is highly saline as a result of being fed by groundwater that flows through subterranean salt deposits formed more than 300 million years ago. Long exploited by humans as a source of salt, especially during the Civil War, the marsh is highly disturbed. Interestingly, some of the plants present are disjuncts from coastal salt marshes on the coast of the Atlantic Ocean. Among these are saltmarsh bulrush (*Bolboschoenus robustus*) and blackgrass rush (*Juncus gerardii*), which are dominant plants in this rare habitat.

6.3.4 Linear Riparian Zones

Linear non-forested riparian zones are found along streams in both the Appalachians and the Ozarks (Fig. 6.8). In the Appalachians, small high-gradient headwater streams often occur beneath a forest canopy and thus their riparian zones would not

Fig. 6.8 Linear streamside habitat in the Ozarks of northern Arkansas. (Photo by Karen Willard)

be considered as non-forested habitats. In both regions, larger streams have riparian zones that can extend out beyond the stream itself for several meters or more. The vegetation of this streamside riparian habitat is very different from that of the adjacent forest. One factor that accounts for this difference is disturbance. Fluctuating water levels, the scouring effects of extreme high-water events, and the periodic deposition of large amounts of debris as result of the latter tend to prevent the establishment of woody vegetation. The actual substrates available for plant establishment range from sand bars to gravel and rocks of various sizes to areas of exposed bedrock. The trees most likely to be found within or adjacent to this habitat are box elder (*Acer negundo*) and sycamore (*Platanus occidentalis*), while various species of willows (*Salix* spp.) occur as small trees. Despite the disturbance factor, riparian vegetation can be fairly biodiverse, with numerous species of herbaceous plants present. Because streams represent ecological corridors through regions of otherwise forested landscape, they sometimes allow the migration of non-native invasive plants into new areas to take place.

6.3.5 Springs

Natural springs, places where groundwater emerges to the surface and flows across the ground, are a common feature of the landscape throughout the forested areas of the Appalachians and Ozarks. Spring-fed streams create unique biotic communities of all types and represent important habitats for both plants and animals. The three largest single-conduit springs in the United States are found in the Ozarks, and the associated forests near the streams include trees of impressive productivity and size since these tend to represent protected areas.

6.4 Rocky Non-forested Habitats

The mountains of the Appalachians are mostly capped with sandstone, and the latter is often exposed on summits and ridges as large tilted blocks or boulder fields (Fig. 6.9). These rock outcrops create a habitat in which the organisms present are exposed to high solar radiation, and during the summer. This results in temperatures appreciably higher than for surrounding areas. This is a very xeric habitat, and the thin soils are largely restricted to crevices and spaces among boulders.

6.4.1 Rock Outcrops

Like bogs, rock outcrops represent a habitat that is often dominated by nonvascular plants. Lichens, many of them crustose, are abundant on the upper surfaces of rocks. Among the more common species in the Appalachians are the common toadskin lichens (*Lasallia papulosa* and *L. pensylvanica*), golden moonglow lichen

Fig. 6.9 Rock outcrop on a mountain ridge in the Ridge and Valley of southwest Virginia. (Photo by Steven L. Stephenson)

Fig. 6.10 Minnie-bush (*Rhodendron pilosum*) is characteristically associated with rock outcrops in the Appalachians. (Photo by Steven L. Stephenson)

(*Dimelaena oreina*), toadstool lichen (*Melanelia culbersonii*), smooth rock tripe (*Umbilicaria mammulata*), and salted shield lichen (*Parmelia saxatilis*). Where a thin layer of soil has developed, mosses occur on the same rocks. Although rock outcrops are classified as a non-forested habitat, an occasional tree (often yellow birch [*Betula alleghaniensis*] or black birch [*Betula lenta*]) is sometimes present. Larger crevices among the rocks provide enough space for shrubs such as minnie-bush (Fig. 6.10) to become established, and ferns are characteristically present and often abundant in favorable microsites.

Rock outcrops are less extensive in the Ozarks and are more likely to take the form of bluffs or cliff faces than tilted blocks or boulder fields. Because these outcrops tend occur at lower elevations than in the Appalachians, they often present much less severe environmental conditions for the organisms present. As a result, they support many of the same plants found in adjacent communities, although lichens and mosses are common to abundant on actual rock surfaces. Areas of exposed rock are very often associated with the glade habitats mentioned earlier in this chapter.

6.4.2 Vertical Cliff Faces

Vertical cliff faces are found in both regions (Fig. 6.11). Examples include the New River Gorge and Linville Gorge in the Appalachians along with the Grand Bluffs Natural Area and portions of the Buffalo River, Jacks Fork River, and Current River in the Ozarks, specifically in the Ozark National Scenic Riverways. Although limited in areal extent, this habitat often supports distinctive plant communities. Vegetation is usually rather sparse and dominated by non-vascular plants, mostly lichens and mosses. However, vascular plants occur where soil has accumulated in crevices and small ledges. Species characteristically associated with cliff faces include littleflower alum root (*Heuchera parviflora*) and roundleaf catchfly (*Silene rotundifolia*). Rock-inhabiting ferns are also characteristic, and several species are

Fig. 6.11 Cliff face on Red Rock Mountain in the Ridge and Valley of the Appalachians. (Photo by Gary P. Fleming)

shared by the two regions. Among these, Bradley's spleenwort (*Asplenium bradleyi*) and lobed spleenwort (*Asplenium pinnatifidum*) occur on acidic-rock cliffs, while walking fern (*Asplenium rhizophyllum*) and bulblet bladder fern (*Cystopteris bulbifera*) occur on calcareous, limestone and dolostone cliffs. There are stunted and often gnarled trees present in this habitat. Arkansas alumroot (*Heuchera villosa* var. *arkansana*) is found on sandstone crevices in the Ouachita Mountains and the Boston Mountains.

On sandstone and igneous cliffs, blackjack oak (*Q. marilandica*), shortleaf pine, and black hickory (*Carya texana*) are likely to be present. Species of trees found along dolomitic and limestone cliff faces include chinquapin oak, gum bumelia, and blue ash (*Fraxinus quadrangulata*) as well as eastern red cedar. Some of the trees (e.g., eastern red cedar in both the Appalachians and Ozarks), because of their inaccessibility and slow growth, are of considerable age. Indeed, the oldest trees in the Ozarks are eastern red cedar and Ashe's juniper (*Juniperus ashei*) in part owing to the inaccessibility but also to the slow growth of these two species. These two species cover a large area along the cliff face of the White River along the Missouri/Arkansas border.

6.4.3 Bolder Streams and Boulder Fields

There is one other special habitat that should be mentioned, albeit one that is restricted almost entirely to certain areas of the Appalachians. Large boulders often occur on the crests of ridges, upper slopes, and below exposures of caprocks in the Appalachians. These boulders are derived from the resistant sandstone that caps the ridges. When the crest is relatively level, boulder-strewn areas sometimes occur. In some localities, especially on the most exposed ridges, these are non-forested habitats, but in other less exposed localities, the boulder-strewn areas can be colonized by at least some trees (Fig. 6.12). The latter occur in places where there is enough

Fig. 6.12 Bolder field on Massanutten Mountain in the Appalachians. (Photo by Gary P Fleming)

space among the boulders for them to become established over time. However, on the side slopes of the ridges, boulders become concentrated along the floors of streamless small valleys (locally called "hollows"). These concentrated areas of boulders are called boulder streams or rock streams (Mills and Stephenson 1999). Except for very large boulder streams, these are not necessarily non-forested habitats, since a boulder stream can be colonized by trees in the same manner as mentioned above. The presence of the trees and the shade they create limit the occurrence of lichens on the boulders, and except for certain vines such as Virginia creeper (*Parthenocissus quinquefolia*), there are few other plants commonly present. Boulder streams are common in the Appalachians, especially in the Ridge and Valley Province. When traveling along a road that was constructed across the side of a slope, it is not uncommon to observe a boulder stream in one or more the hollows visible upslope from the road.

Fig. 6.13 Rocky glade in the Ozarks of northern Arkansas. (Photo by Steven L. Stephenson)

6.4.4 Gravel Glades

Gravel or gravel-rock glades (*sensu* Quarterman et al. 1993) are found in some areas of the Ozarks (Fig. 6.13). These are characterized by their very thin, often gravel covered soil, and with some areas consisting of bare rock. Gravel glades are the result of the presence of what are called chert nodules. Chert nodules occur as inclusions in a type of limestone or dolomite characteristic of what geologists refer to the Boone Formation, which is exposed at the surface at a number of places on the Ozark Plateaus. When the sediments that formed the rocks of the Boone Formation were deposited during the period of about 325–360 million years ago, some calcium carbonate (the primary component of limestone) was replaced by silicon dioxide. This produced chert nodules, which break up rather easily, ultimately producing gravel-sized particles. The gravel found in stream beds in the Ozarks has a totally different origin and should not be confused with the gravel of gravel glades. This is an extremely xeric habitat during summer and is inhospitable for all but the most tolerant plants, including foliose lichens such as the earth lichen (*Dermatocarpon lachneum*). Soils are never deep enough to allow the establishment of shrubs and trees. It should be noted that some ecologists prefer the term "barrens" when referring to open, rocky habitats. This certainly would apply to gravel or gravel-rock glades.

References

Bartgis RL (1985) Limestone glade in West Virginia. Bartonia 51:34–36

Bartgis RL (1993) The limestone glades and barrens of West Virginia. Castanea 58:69–89

Baskin JM, Baskin CC (2000) Vegetation of limestone and dolomite glades in the Ozarks and Midwest region of the United States. Ann MO Bot Gard 87:286–294

Baskin JM, Baskin CC (2004) History of the use of "cedar glades" and other descriptive terms for vegetation on rocky limestone soils in the Central Basin of Tennessee. Bot Rev 70:403–424

Clarkson RB (1964) Tumult on the mountains. McClain Printing Company, Parsons, WV

Core EL (1966) Vegetation of West Virginia. McClain Printing Company, Parsons, WV
Davis MB (1983) Holocene vegetation history of eastern United States. In: Wright HE (ed) The late-quaternary of the United States, vol 2. University of Minnesota Press, Minneapolis, pp 166–181
Fortney RH (1993) Canaan Valley—an area of special interest within the upland forest region. In: Stephenson SL (ed) Upland forests of West Virginia. McClain Printing Company, Parsons, WV, pp 47–65
Francl KE, Ford WM, Castleberry L (2004) Characterization of high elevation central Appalachian wetlands, United States Department of Agriculture, Forest Service Research Paper NE-725. United States Department of Agriculture, Newtown Square, PA
Gentry JL, Johnson GP, Baker BT, Witsell CT, Ogle JD (2013) Atlas of the vascular plants of Arkansas. PMC Solutions, Fayetteville, AR
Jeffries DL (1985) Analysis of the vegetation and soils of glades on Calico Rock sandstone in northern Arkansas. Bull Torrey Bot Club 112:70–73
Jeffries DL (1987) Vegetation analysis of sandstone glades in Devil's Den State Park, Arkansas. Castanea 52:9–15
Mark AF (1958) The ecology of the southern Appalachian grass balds. Ecol Monogr 28:293–336
Mills HH, Stephenson SL (1999) Forest vegetation and boulder streams in the central Appalachian Valley and Ridge Province, southwestern Virginia. Bull Torrey Bot Soc 126:188–196
Nelson PW (2010) The terrestrial natural communities of Missouri. Missouri Department of Natural Resources, Jefferson City, MO
Pittillo JD (1994) Vegetation of three high elevation southern Appalachian bogs and implications of their vegetational history. In: Trettin CC, Aust WN, Wisniewski J (eds) Wetlands of the interior Southeastern United States. Springer, Dordrecht, pp 333–248
Platt RB (1951) An ecological study of the mid-Appalachian shale barrens and of the plants endemic to them. Ecol Monogr 21:269–300
Quarterman E, Burbanch MB, Shure JD (1993) Rock outcrop communities: limestone, sandstone, and granite. In: Martin WH, Boyce SG, Echternacht AC (eds) Biodiversity of the southeastern United States: upland terrestrial communities. John Wiley and Sons, New York, pp 35–86
Rentch JS, Fortney RH (1997) The vegetation of West Virginia grass bald communities. Castanea 62:147–160
Schultheis EH, Hopfenspereger KN, Brenner JC (2010) Potential impacts of climate change on sphagnum bogs of the southern Appalachian Mountains. Nat Areas J 30:417–424
Ware S (2002) Rock outcrop plant communities (glades) in the Ozarks: a synthesis. Southwest Nat 47:585–597
Weigl PD, Knowles TW (1995) Megaherbivores and southern Appalachian grass balds. Gowth Change 26:365–382
Wieder RK, McCormick AM, Lang GE (1981) Vegetational analysis of big run bog, a nonglaciated sphagnum bog in West Virginia. Castanea 46:16–29

Chapter 7
Animals of the Appalachians and Ozarks

Abstract The forests of the Appalachians and Ozarks are home to a diverse assemblage of animals. With some exceptions, the mammals, birds, reptiles, and amphibians found in the Appalachians also occur in the Ozarks. The exceptions are species associated with the high-elevation broadleaf and coniferous forests of the Appalachians, which have no vegetative counterpart in the Ozarks; species restricted to special habitats in one of the two regions; or species with a distribution centered in western North America but also occur in the Ozarks. The purpose of this chapter is to describe many of the more characteristic terrestrial or largely terrestrial animals of the Appalachians and Ozarks. Unique aquatic examples also are considered.

Keywords Terrestrial animals · Birds · Neotropical migrants · Mammals · Reptiles · Salamanders · Distribution patterns · Carnivores · Herbivores

7.1 Mammals

7.1.1 Large Mammals

The very largest animals are by far the best known, and all of these are mammals (Sealander and Heidt 1990; Linzey 2021). The white-tailed deer (*Odocoileus virginianus*) is the most common large mammal in both regions (Fig. 7.1). Adults generally weigh between 60 and 130 kg. The upper body of an adult deer is reddish brown in spring and summer and grayish brown during fall and winter, with a white lower body and throat. Deer are browsers, feeding upon the emerging leaves of woody twigs they can reach. In broadleaf forests where the deer population may be high, a "browse line" can be observed that represents the height beyond which they could not reach. Deer also feed upon low-growing plants in a forest along with

S. L. Stephenson, R.-M. Muzika, *Forests of the Southern Appalachian Highlands and Interior Highlands of the United States*, Plant and Vegetation 20,
https://doi.org/10.1007/978-3-032-01155-8_7

Fig. 7.1 The white-tailed deer (*Odocoileus virginianus*) is the most common large mammal in both the Appalachians and Ozarks. (Photo by Steven L. Stephenson)

acorns, lichens, and mushrooms. During the peak season for mushrooms that occurs in summer and fall, they appear to be an "item of choice" in the diet of deer (Stephenson 2010). Unfortunately, deer sometimes become so acclimated to humans that they venture into urban and suburban areas, where they feed upon ornamental plants and garden crops, causing considerable damage.

There is little question that the white-tailed deer can have a major and largely detrimental effect on understory vegetation, and this includes more than woody plants (Miller et al. 2009). Deer feed upon spring ephemeral plants, and over a period of years this can reduce the population size of most species. In contrast, unless there are no other sources of food present, deer do not ordinarily feed upon ferns. Because populations of the white-tailed deer are lower in areas of the Ozarks than they are in the Appalachians, their effect upon the vegetation is much less evident, although still present. The dependence of the white-tailed deer on acorns cannot be overstated. Acorn abundance varies from year to year. When acorns are abundant, they dominate the diet of the white-tailed deer. When they are not, deer feed mostly upon the leaves of plants. Interestingly, Wentworth et al. (1990) reported that the scarcity of acorns significantly lowered the reproductive rates of yearling white-tailed deer, although older individuals were not affected.

The eastern elk (*Cervus canadensis*) once occurred throughout the Appalachians and Ozarks but were hunted to extinction in the mid-nineteenth century, with the last known elk in either region killed in 1875 in West Virginia. Elk are large animals, with males exceeding 400 kg and standing 1.5 m at the shoulder. The fur of an elk is tawny brown, but the most distinctive feature is the presence of large antlers on the male. Elk have a diet similar to that of the white-tailed deer and were associated with open areas in the forest. Efforts to reintroduce the elk to some portions of its former range have had mixed success, but some introduced populations have thrived. Most reintroductions in the Appalachians have taken place on the Allegheny and Cumberland Plateaus in the states of Kentucky, Tennessee, Virginia, and West Virginia. The reintroduction in Kentucky has been particularly successful, with a current population of about 15,000 individuals. Elk have demonstrated a capability to spread well beyond the site where they were introduced (Lituma et al. 2021). Elk were likely extirpated from much of the Ozarks soon after the civil war. Elk were successfully reintroduced in the Ozarks of Arkansas in the 1980s, with the notable example being the herd that now exists on the Buffalo National River in Arkansas and in the Current River Hills of the Ozarks of Missouri, where they were reintroduced in the early 2000s and currently thrive.

A much larger mammal—the American buffalo (*Bison bison*)—occurred in the Appalachians in the recent past. Adult males sometimes exceeded a weight of 900 kg. Some people think that the buffalo found in the Appalachian region was a smaller variant (known as the "woodland buffalo") of the one that was so abundant in the prairies of the western central United States. The numbers of the "eastern buffalo" never approached those found in the grasslands of the western central United States, but there were enough to be hunted in the seventeenth and early eighteenth centuries. It seems likely that the presence of the buffalo in the east indicates that there may have been more savanna and grassland habitat available than generally thought to be the case. The last known buffalo (a cow and her calf) in the Appalachians was killed in 1825 in Randolph Country, West Virginia. The buffalo was more common in the Ozarks, especially on the western boundary where the vegetation of the region transitions from forests to grassland. The pattern of bison distribution in the Ozarks was linked to a complex relationship that existed among fire, settlement, and available bison habitat (Nanavati and Grimm 2020). This animal was killed off in the early 1800s, but small herds have been reintroduced to the region. In both the Appalachians and the Ozarks, the buffalo left a legacy of place names, such as the Buffalo River in Arkansas and Buffalo Creek in Virginia.

In the more distant past, there were much larger animals in the Appalachians and Ozarks (Guilday 1982). During the late Pleistocene (ca. 12,000 years ago), the American mastodon (*Mammut americanum*) occurred in both regions. This animal stood almost 3 m high at the shoulder and could weigh 7000 kg. Other large mammals of the late Pleistocene whose fossils have been found in the two regions are the Jefferson's ground sloth (*Megalonyx jeffersonii*), the giant bison (*Bison latifrons*), and the short-faced bear (*Arctodus simus*), all of which could weigh in excess of 900 kg. It is possible that the Jefferson's ground sloth and giant bison, through their feeding activities, may have been responsible for helping maintain grassy areas in the landscape in which they lived.

All of the mammals mentioned above, with the possible exception of the extinct short-faced bear, are (or were) herbivores and feed upon plants. There are other mammals that are omnivores and thus feed on plants, animals, and a variety of food resources. The largest of these is the black bear (*Ursus americanus*), which has an adult weight of about 160 kg, with males generally larger than females (Fig. 7.2). The black bear has a short tail and, as its name suggests, black fur. Black bears are active both at night and during the daylight hours. Although considered as omnivorous, a major portion of a bear's diet is derived from plant parts, including foliage, fruits, and roots. They also feed upon insects and small vertebrates. When they are available, the fawns of the white-tailed deer are heavily preyed upon by black bears.

Female black bears typically breed every other year in mid-June to July and produce a litter of two to four cubs in the den in late January to early February. The cubs

Fig. 7.2 The black bear (*Ursus americanus*) is an omnivore and feeds upon both plants and animals throughout both the Appalachians and Ozarks. (Photo by Tyler Reber)

are very immature when they are born, lacking hair and weighing less than a pound. They typically first leave the den in early April but do not disperse until their second spring. Black bears, with the exception of a female with cubs, tend to be rather solitary animals. Because of their size, they are potentially dangerous to humans, but most encounters between humans and bears result in the latter running away. However, it is best to maintain a safe distance from a female with cubs.

Black bears have a wide range in North America, but populations have fluctuated over time. Black bears were extirpated from the Ozarks but reintroduced in 1958 to Arkansas from Minnesota (Smith and Clark 1994). Only recently have black bears expanded into the forests of Missouri, having dispersed from populations in the Arkansas Ozarks. Populations of the black bear are now larger in the Appalachians and Ozarks than they have been since prior to the arrival of Europeans. Early settlers hunted the black bear to such an extent that the animal became relatively rare.

Extensive logging in the two regions and the loss of American chestnut in the Appalachians (which reduced their food resources and changed forest structure and composition) also had a negative effect on populations. However, since large-scale logging ended, populations have continued to climb and the black bear has become relatively common. Reintroductions of the black bear to some areas of the Ozarks have also taken place. Obviously, the increased numbers enhance the chances of a human-bear encounter.

Hogs (*Sus scrofa*) are not native to either the Appalachians or Ozarks but were introduced by early European settlers in the two regions. Some escaped and formed feral populations. These have occurred in both regions for more than two centuries and are exceedingly well established. Moreover, hogs still occasionally escape farms, to augment the existing wild, free-roaming population. There are both Eurasian wild hogs and hybrids. Feral hogs are a serious issue in the forests of both the Ozarks the Appalachians. These animals are discussed in more detail elsewhere in this book.

7.1.2 Medium-Sized Mammals

The eastern cottontail (*Sylvilagus floridanus*) is not an inhabitant of closed canopy forests in the Appalachians and Ozarks but can occur in open, early successional forests with a dense ground cover. However, grassy glades and meadows are its usual habitat. A tail that resembles a cotton ball, the prominent ears, and large hind feet are the distinguishing features of this animal. Cottontail rabbits, which weigh between 0.8 and 2.0 kg as adults, feed on grasses and the stems and leaves of whatever other herbaceous plants are available.

The Appalachian cottontail (*Sylvilagus obscurus*) is a smaller (0.75–1.2 kg) and much less common rabbit found at higher elevations (generally >600 cm) in the mountains of the Appalachians. The Appalachian cottontail is not easily distinguished from the eastern cottontail, although it tends to be lighter in color and has relatively shorter ears. The different habitats in which the two species occur is probably the most important way to tell them apart.

The preferred habitat of the groundhog (*Marmota monax*) is the edge of a forest, although they sometimes occur in open grassy areas and occasionally can be found within a forest. This animal is a member of the same family (Sciuridae) as squirrels and would be considered as a ground squirrel. Groundhogs are large stout-bodied animals and can weigh up to 6 kg. Their thick fur is some shade of brown, their legs are short, and they have a short, bushy tail. They feed upon all types of leafy plants but also can consume small woody twigs and bark.

The beaver (*Castor canadensis*) is one of the larger mammals (with weights up to 50 kg) found in the Appalachians and Ozarks. Beavers are rodents and thus related to mice, rats, squirrels, and groundhogs. However, beavers are much larger than any of these. In fact, beavers are the second-largest living rodents. Interestingly, during the Pleistocene, the now extinct giant beaver (*Castoroides ohioensis*) was more than twice as large. Fossils of the giant beaver are known from both the Appalachians and Ozarks. Beavers are semiaquatic mammals associated with all types of freshwater habitats, including rivers, small streams, lakes, and ponds. They have stout bodies with large heads, long chisel-like front teeth, brown or gray fur, somewhat hand-like front feet, webbed rear feet, and tails that are flat and scaly. They are herbivores, feeding upon tree bark, grasses, sedges, and various aquatic plants.

Beavers are distinct because of their unique behavior and what they accomplish as a result. They build dams that impede the flow of water and lodges in which they live. Both are constructed using tree branches, other types of woody debris, rocks, and mud. They actually fell trees for building materials, and sometimes the trees can be quite large. Their construction of a dam creates a wetland that provides a habitat for numerous other species. When they abandon a particular site because the food resources are exhausted, beavers move upstream or downstream and build another dam. Over time, this creates a mosaic of different aged emergent to woody wetlands along a particular stream. A good example of this situation can be found in Canaan Valley in West Virginia. Beavers are considered as a keystone species or the ultimate ecological engineer in the wetlands they create. Beavers are absent from the small, high-gradient streams of that exist at higher elevations in the Appalachians but do occur in wetland areas that occur at the same elevations.

The beaver, otter, mink, and muskrat are semiaquatic mammals found throughout the forests of the Appalachians and Ozarks. Beavers were among the most numerous mammals in North America before the first Europeans arrived, with an estimated 100–200 million animals (Lituma et al. 2021). Beginning as early as the mid-1600s, Native Americans began trading beaver pelts to Europeans. The Europeans themselves began killing beavers for their pelts about the same time and this quickly intensified. This soon led to a dramatic decrease in the populations of beavers in both the Appalachians and Ozarks during the 1700s and 1800s. By 1900, the species had been almost entirely eliminated over much of its former range (Smith 1993). A change in clothing fashions ultimately caused the market for beaver pelts to drop, and surviving populations began to recover. However, beavers had to be reintroduced to many places where they once occurred.

One relatively common but not often observed medium-sized mammal in the Appalachians and Ozarks is the raccoon (*Procyon lotor*), easily recognized by the black band over its face and the bushy black tail. Racoons typically weigh from 4 to 9 kg and are omnivorous, feeding on food derived from both plants and animals. They are nocturnal and observed less frequently than might be expected for an animal of their size.

The striped skunk (*Mephitis mephitis*) is somewhat smaller than the racoon, usually weighing between 1.5 and 4.5 kg. Like the chipmunk (described later in this chapter), the striped skunk is easily recognized. The two broad white stripes across the back that contrast with the otherwise white fur are distinctive. Striped skunks are nocturnal, where they seek out such things as earthworms and grubs for food. As is generally known, skunks have well-developed scent glands. The odor they produce serves as the most effective defense of any animal found in the Appalachians or Ozarks.

The eastern spotted skunk (*Spilogale putorius*) is much smaller (typically weighing 0.2–1.8 kg) and has a more slender body than the striped skunk. It can be distinguished from the latter by the presence of four broken white stripes on the back that give the animal a "spotted" appearance. The eastern spotted skunk is much less common than the striped skunk and should be considered as rare in some portions of its range. Like the latter, it is nocturnal and omnivorous. The usual habitat is a shrubby, often rocky area at lower elevations in the Appalachians, but individuals are occasionally spotted at high elevations. Populations of the eastern spotted skunk have declined dramatically over the past century, for reasons that remain unknown (Thorne et al. 2024).

The eastern spotted skunk is not found in the Ozarks. However, the plains spotted skunk, *Spilogale interrupta* (formerly *Spilogale putorius interrupta*), is found in the Ozarks of Missouri and Arkansas, where this species is considered as endangered. In other areas of the Midwest, these animals typically occur in cultivated lands or prairies, but in the Ozarks, they are associated with woodland habitats with extensive leaf litter and fallen logs.

The Virginia opossum (*Didelphis virginiana*), the only marsupial found in North America, is common in both the Appalachians and Ozarks. The ancestor of the Virginia opossum migrated north from South America some 3 million years ago. This is another intermediate-sized mammal, typically weighing between 2 and 6 kg. The elongated snout, grayish-white fur, and the naked, scaly tail distinguish the opossum from any other animal one is likely to encounter. Opossums are most abundant at lower elevations in the Appalachians but common throughout the Ozarks. These animals feed upon virtually anything, including carrion.

The most unusual medium-sized mammal found in the Ozarks is the nine-banded armadillo (*Dasypus novemcinctus*), a species native to Mexico but which has been expanding its range for over a century through both natural dispersion and intentional introduction. It may have appeared in the Ozarks as recently as 50 years ago. These animals can now be found in the northern Ozarks but they are limited by cold temperatures. More recently, there have been numerous reports of armadillos in the Appalachians, which indicates that the animal's range has continued to expand. The animal is more common on the Cumberland Plateau than elsewhere in the region. The armadillo is easily recognized by its segmented, shell-like body covering. Adult animals can reach a weight of up to 20 kg. Armadillos feed primarily on insects and other invertebrates and are most active at night.

Woodrats are medium-sized (250–480 g) mammals found in both the Appalachians and Ozarks, although two different species are involved. The eastern woodrat (*Neotoma floridana*) is found in the Ozarks, whereas the Allegheny woodrat (*N. magister*) occurs in the Appalachians. These animals, which inhabit a variety of different types of habitats, are best known for the very large (sometimes as much as a meter across) nests they constructs out of twigs, other types of plant debris, and anything else that is available. This fact accounts for the woodrat's other common name of pack rat. Woodrats store food caches in the nest that are used during the winter months. The Allegheny woodrat occurs along the Appalachian Mountains with a relatively widespread distribution. Populations have declined notably in some areas. The animal's habitat includes cliff edges and rocky habitats. The woodrat should not be confused with the Norway rat (*Rattus norvegicus*), an introduced pest species.

7.1.3 Small Mammals

With few exceptions, including the examples noted above, most of the mammals found in the Appalachians and Ozarks are small and inconspicuous. However, there are a few that are both common and active enough to be well known to most people. This is certainly the case for the gray squirrel (*Sciurus carolinensis*). In addition to being a common inhabitant of forests, gray squirrels have become well adapted to urban areas and are often common in city parks. Gray squirrels can weigh 500 g or even more. They feed mostly on nuts and other types of plant parts but occasionally consume small animals such as insects.

There are at least seven species of squirrels in the Ozarks, with the flying squirrels of particular interest. The flying squirrels (*Glaucomys* spp.) are the most unusual small mammals (typical weight of 70–130 g) found in the Appalachians and Ozarks. The presence of a fur-covered membrane that extends from the wrist to the ankle of the legs and a broad flat tail enables the flying squirrel to glide through the air as it descends from a tree or branch. These animals are nocturnal and uncommon enough in those areas in which they occur to be rarely seen. The northern flying squirrel (*Glaucomys sabrinus*) is truly uncommon and limited to high-elevation and typically coniferous forests, whereas the southern flying squirrel (*G. volans*) is common in forests at lower elevations and also is found in the Ozarks. The diet of the southern flying squirrel consists mostly of plant material (especially acorns, other nuts and fruits), whereas the northern flying feeds mostly on fungi and lichens, the latter to a much greater extent than any other animal in the entire region of the Appalachians and Ozarks.

Perhaps the most noticeable small mammal in the two regions is the eastern chipmunk (*Tamias striatus*), which is easily recognized from its distinctive coloration. Chipmunks have reddish-brown fur on their upper body, with five dark brown stripes and contrasting light brown stripes along their backs. These animals are most abundant in oak forests, where their populations are known to fluctuate widely from year to year. The explanation for this phenomenon is described below. Chipmunks are omnivorous, feeding upon acorns, seeds, berries, insects, and mushrooms.

Oaks have long been known for having "mast years" when they produce a huge number of acorns in a particular year, far more than the year before or the year after. This is a major ecological advantage for the oak trees because they overwhelm the animals that feed upon their acorns. With an abundance of acorns, seed-eating animals are likely to be satiated long before they eat all the acorns produced in a mast year, leaving the rest to sprout and produce seedings. In a non-mast year, it is possible that most of the acorns could be consumed. The huge number of acorns enhances the chances of oaks having their acorns carried away from the parent tree and buried by animals such as squirrels. If the squirrel doesn't dig up an acorn, it is already in the soil and stands a good change of germinating and giving rise to an oak seeding. Moreover, if the squirrel carries the acorn some distance away from the parent tree, short distance dispersal is accomplished.

The most abundant mammals in the forests of the Appalachians and Ozarks also are the very smallest. This group consists of the deer mice, voles, and shrews. Of these, deer mice are undoubtedly the most numerous (Fig. 7.3). There are two species, the white-footed deer mouse (*P. leucopus*) and the eastern deer mouse (*Peromyscus maniculatus*). These two species are morphologically very similar, with the length and color of the tail representing the primary distinguishing feature. Both are small, with an average body weight of about 21 g. Deer mice are most common in forests with oaks as a major component (Healey and Brooks 1988).

The meadow jumping mouse (*Zapus hudsonius*) occurs throughout the Appalachians but is limited to only the northernmost portion of the Ozarks. This is another small mammal (12–25 g) and perhaps the most distinctive. The meadow jumping mouse can be easily recognized from its large hind feet, very long tail, and

Fig. 7.3 The deer mice, including the white-footed deer mouse (*Peromyscus leucopus*) and the eastern deer mouse (*P. maniculatus*), are among the most common small mammals in the Appalachians and Ozarks. (Photo by Chris Helzer)

relatively short front limbs. This species is associated with moist, non-forested habitats. Its most remarkable feature is the ability to jump distances of a half meter or more.

The woodland vole (*Microtus pinetorum*) is another of the small mammals found in the forests of the two regions. It is similar in appearance to the deer mouse but larger (14–37 g), fuller-bodied, and has a much shorter tail. The woodland vole (or sometimes called the pine vole) is associated with the litter layer on the forest floor, where it feeds on all types of plant material and insects. The rock vole (*Microtus chrotorrhinus*), so named because of the rocky slopes that represent its usual habitat, is an example of a small mammal restricted to high-elevation areas of the Appalachians and not found in the Ozarks. The rock vole is slightly larger (30–48 g) than the woodland vole but has a similar diet. The eastern red-backed vole (*Clethrionomus gapperi*), which can be distinguished from all other voles by its reddish back, occurs only in the Appalachians. This is a very small vole, sometimes weighing less than 10 g.

The northern short-tailed shrew (*Blarina brevicauda*) is another small mammals (with a body weight of 15–30 g) found in the forests of the Appalachians and Ozarks. All shrews are very secretive animals rarely observed in nature. The northern short-tailed shrew spends most of its time in shallow burrows beneath the litter layer, where it feeds upon insects, other invertebrates, and fungi. This animal can be recognized from its long-pointed shout, short tail, and very small eyes and ears. Because of its very high rate of metabolism, the northern short-tailed shrew, like all shrews, remains almost constantly active in its search for food. The southern short-tailed shrew (*B. carolinensis*) also occurs in the Appalachians and Ozarks, although it has a limited distribution in the former region. This shrew is very small, usually weighing less than 14 g. Diversity of shrews is higher in the Appalachians than Ozarks, at least in part due to the greater diversity of habitats available. Shrews have two unusual features not shared with other small terrestrial mammals in the Appalachians and Ozarks. First, some shrews, including the northern short-tailed shrew, are venomous. Second, like bats, shrews use echolocation to compensate for their exceedingly poor vision.

7.1.4 Carnivores

Smaller animals (and sometimes even large animals) are a source of food for the carnivores found in the Appalachians and Ozarks. The mountain lion (*Puma concolor*) was once found throughout the Appalachians and Ozarks. It has now been considered extinct for more than a century, although there continue to be occasional "sightings" of this animal in the mountainous, less populated areas of Virginia and West Virginia. Numerous but widely scattered sighting also have been reported for the Ozarks. However, these individuals may have been trapped elsewhere and released. The mountain lion is thought to have been eliminated from the Ozarks in the early twentieth century, but there have been confirmed reports of its occurrence in some portions of the Missouri Ozarks falling. There is even some evidence that a breeding population of mountain lions might occur in the Ozarks.

Although rarely observed in nature, the bobcat (*Lynx rufus*) is not uncommon in the Appalachians and Ozarks. Like the mountain lion, the bobcat is carnivorous, feeding mostly on smaller mammals such as rabbits, squirrels, voles, and mice along with small birds. Bobcats are much smaller than mountain lions, usually weighing no more than 9–12 kg.

Two types of foxes occur in Appalachians and Ozarks. These are the gray fox (*Urocyon cinereoargenteus*) and the red fox (*Vulpes vulpes*). Although similar in general appearance, they are members of different genera and thus not closely related. The red fox is generally larger (3–10 kg) than the gray fox (3–6 kg). Color variations are encountered in both species, but the fur of the red fox is typically red while that of the gray fox is silvery gray. Red foxes are more common in mixed forests and relatively open areas, whereas gray foxes are largely restricted to forested habitats. Over the past century, gray foxes have been becoming increasingly less common and have been extirpated from some areas of their natural range. Both species co-occur in the same forest habitats, but the red fox is much more common. The two types of foxes are preferentially carnivorous by choice but also feed upon nuts and berries. The gray fox is capable of climbing trees, which gives it a wider choice of prey.

The coyote (*Canis latrans*), a third member of the dog family (Canidae), is native to the Ozarks but has spread to the Appalachians. This is a larger animal than either of the foxes, with adult males often weighing more than 15 kg. The fur is usually grayish-brown but can be black. Because of their larger size, coyotes are capable of taking down larger prey than foxes. For example, they are capable of taking down a white-tailed deer (especially a fawn) and readily prey upon turkeys (*Meleagris gallopavo*). Coyotes prefer open habitats and are not forest animals except when they prey on animals typically found in the forest interior.

The red wolf (*Canis rufus*) once occurred throughout the southeastern United States but was essentially eliminated by the mid-1900s. It was declared globally extinct in 1980. However, red wolf-coyote hybrids can be found in the region, and efforts have been made to reintroduce this animal to portions of its original range. There is some question as to whether the red wolf ever occurred in the upland areas of the Appalachians and Ozarks or was confined to lowland more coastal areas outside of these two region. However, in 1950 the last red wolf in the state of Missouri was killed in the southwestern part of the state.

There are two small, slender-bodied, short-legged, and extremely active carnivores found in the Appalachians and Ozarks. These are the long-tailed weasel (*Mustela frenata*) and the American mink (*Neogale vison*). The weasel (85–260 g) is smaller than the mink (600–900 g) and totally terrestrial whereas the mink is semiaquatic. The latter actively hunts in water and can dive well below the surface. Weasels feed mostly on small animals and birds, and they are capable of killing animals larger than themselves. As is widely known, the fur of the mink is highly prized for its use in clothing. Both the weasel and mink are secretive animals and may occur in an area without this being known.

The river otter (*Lontra canadensis*) is a much larger (5–14 kg) semiaquatic carnivore found in the two regions. This animal's distinctive features are the flattened head, long whiskers, webbed feet, and long tapered tail. Otters are the top predator in the river food chain and feed mostly on fish, crawfish, mussels, and frogs. Intensive hunting and trapping during the 1800s and 1900s extirpated the otter from portions of its natural range, but recent reintroductions have allowed populations of this animal to be restored.

7.1.5 Bats

The forests of the Appalachians and Ozarks are the home for more than a dozen species of bats, including the big brown bat (*Eptesicus fucus*), the Indiana bat (*Myotis sodalist*), the eastern red bat (*Lasiurus borealis*), evening bat (*Nycticeius homeralis*), northern long-eared bat (*Myotis septentrionalis*), and the tricolored bat (*Perimyotis subflavus)*. These range in size from the eastern red bat (33 cm wingspan) to the tricolored bat (23 cm wingspan). The two most commonly observed species are the big brown bat (Fig. 7.4) and the eastern red bat. During the winter, most bats hibernate in caves or more recently in abandoned mines and other secluded places where they are protected from extreme winter temperatures. The one noteworthy exception in the Appalachians and Ozarks is the eastern red bat, which migrates south during winter. During the summer, most species roost in trees. The endangered Ozark big-eared bat (*Corynorhinus townsendii* subsp. *ingens*) is currently found in north central and northwestern Arkansas and northeastern Oklahoma. This species once occurred in the Missouri Ozarks but has not been observed since 1971 (Sasse 2022). Bats feed on flying insects, ranging in size from midges to large moths. In the past two decades, a fungal disease called the white nose syndrome and caused by the fungus *Pseudogymnoascus destructans* has severely impacted bat populations throughout much of eastern North America.

Fig. 7.4 The big brown bat (*Eptesicus fucus*) is one of the more common bats found in the Ozarks and Appalachians. (Photo by Michael Slay)

7.2 Birds

7.2.1 Neotropical Migrants

Birds are the most visible of the animals that occur in the Appalachians and Ozarks (Neff 1923; Hall 1983; James and Neal 1986). More than 400 species have been recorded, but the number known to breed in the two regions is much lower. Birds can be divided into two main groups based on their distribution throughout the year. The first and larger group is made up of Neotropical migrants. These birds spend much of the year in the Neotropics but migrate north to the temperate zone for 2 or 3 months each year to breed. Because of their bright colors, these birds tend to be very conspicuous. Examples include the white-eyed vireo (*Vireo griseus*), blue-gray gnatcatcher (*Polioptila caerulea*), northern parula (*Setophaga americana*), red-eyed vireo (*Vireo olivaceus),* prothonotary warbler (*Protonotaria citrea*), black-and-white warbler (*Mniotilta varia*), worm-eating warbler (*Helmitherus vermivcorum*), American redstart (*Setophaga ruticilla*), common yellowthroat (*Geothlypis trichas*), wood thrush (*Hylocichia mustelina*), cerulean warbler (*Dendroica curulea*), orange-crown warbler (*Leiothlypis celata*), and yellow warbler (*S. petechia*). All of these birds breed in the Appalachians and Ozarks, but there are a number of other species that breed only in the high-elevation forests of the Appalachians (Hall 1983, 1984a, b). The magnolia warbler (*Setophaga magnolia*), veery (*Catharus fuscescens*), hermit thrush (*C. guttatus*), chestnut-sided warbler (*S. pensyvanica*), and

Fig. 7.5 The chestnut-sided warbler (*Setophaga pensylvanica*) is one of the more common Neotropical migrants in the Appalachians and Ozarks. (Photo by Matt Orsie)

blackburnian warbler (*S. fusca*) fall into this category. Populations of some of the Neotropical migrants mentioned above are declining, with loss of suitable habitat in the regions of South and Center America where they overwinter seeming representing a major factor (Hall 1984a, b). Another factor, which is important for species such as the cerulean warbler, is the lack of areas of mature forests in the Appalachians and Ozarks. Such forests represent its preferred habitat.

Neotropical migrants tend to be brightly colored birds and are often easily identified by their songs even when they are not visible. This is certainly true for one of the more unusual Neotropical migrants—the eastern whip-poor-will (*Antrostomuc vociferus*). Unlike the other examples noted above, the whip-poor-will is nocturnal, and this bird is cryptically and not brightly colored. Indeed, the predominant colors of the eastern whip-poor-will are gray, brown, and black, occurring together in a pattern not unlike the dead leaves in the forest floor, which is where it rests during the daylight hours. Only at dawn and dusk does the whip-poor-will make the classic call that is the source of its common name.

The chestnut-sided warbler (*Setophaga pensylvanica*) is one of the more common and easily identified Neotropical migrants (Fig. 7.5). In the summer, males are unmistakable in appearance. The head has a yellow crown above a white face with black eye stripes, the underside of the body is white and the sides of the body are chestnut-colored (hence the common name). The coloration of the female is essentially a washed-out version of that of the male. In the fall, the plumage of chestnut-sided warblers has more of a dull greenish appearance. This species tends to be associated with disturbed or second-growth forests, which means that it is probably much more abundant today than before the extensive logging that occurred in the Appalachians and Ozarks during the late 1800s and early 1900s.

7.2.2 Permanent Residents

A second group consists of those birds that are permanent residents and remain in the two regions or at least in the continental United States throughout the entire year. Resident species that occur in both the Appalachians include the northern cardinal (*Cardinalis cardinalis*), American crow (*Corvus brachyrhynchos*), blue jay (*Cyanocitta cristata*), American robin (*Turdus migratorius*), tufted titmouse (*Baeolophus bicolor*), mourning dove (*Zenaida macroura*), red-bellied woodpecker (*Melanerpes carolinus*), white-breasted nuthatch (*Sitta carolinensis*), European starling (*Sturnus vulgaris*), and turkey vulture (*Cathartes aura*). The European starling is an invasive first introduced into the United States in the late nineteenth century.

Woodpeckers play an important role in forests, and there are seven to eight species of woodpeckers in the Appalachians and Ozarks, and most of these are year-round residents. Among species found in the Ozarks, the red-cockaded woodpecker (*Picoides borealis*), which was associated with pine forests or pine woodlands, was likely extirpated from Missouri during the time of intensive logging. This woodpecker is still found in the Ozarks of Arkansas and has been reestablished in Ozarks of Missouri.

Based on bird counts, the northern cardinal is the most abundant bird in the Ozarks. With its bright red color, crest on the head, and large pink bill, the male is unmistakable. The female cardinal is duller in color but shares the crest and pink bill. The cardinal is also one of the more common birds in the Appalachians as a whole, along with other birds such as the American robin and blue jay. However, the assemblage of birds characteristic of the high-elevation areas is more diverse. The American robin is one of the most familiar of all birds, and many people regard the arrival of this species in the spring as one of the first signs of spring. This bird was named after the European robin because of its similar reddish-orange breast, but the two species are not closely related. As noted above, the taxonomic name of the American robin is *Turdus migratorius*, with the specific epithet *migratorius* suggesting that the bird is a migratory species. However, this is not necessarily true. Some robins stay in a specific area year round, while others do migrate. This is especially true for those that spend the summer at high latitudes. The distance migrated varies considerably but only rarely does it approach the distances traveled by Neotropical migrants. For the most part, migration in American robins involves short distances and is linked to the availability of food.

Blue jays are among the best known birds in the Appalachians and Ozarks. They are easily recognized by their blue and white color, the black band around the neck, and the crest on the head (Fig. 7.6). Interestingly, they are in the same family as crows and ravens, but this fact is not readily apparent because of unmistakable differences in color (both crows and ravens are uniformly black) and size (the blue jay is appreciably smaller than either of the two other species). The usual habitat for blue jays is a wooded area, and their diet consists primarily of seeds and fruits, although they often feed upon insects and sometimes small mammals and baby

Fig. 7.6 The blue jay (*Cyanocitta cristata*) is one of the more distinctive birds found in the Appalachians and Ozarks. (Photo by Eva Paulhus)

birds. Blue jays are of particular ecologically importance for two reasons. First, they consume the larvae of both spongy moths and tent caterpillars, two major forest pests. Second, among the things they feed upon are acorns, some of which they bury in the ground but then do not retrieve. Because the acorns can be transported a considerable distance from the tree where they were collected, this greatly increases the dispersal potential of oaks in the forests of the Appalachians and Ozarks. Blue jays are known to be unusually intelligent. For example, they often sound an "alarm" call that alerts other animals to possible danger. They are also rather aggressive birds, especially in defense of their nests.

Birds vary considerably with respect to what they feed upon and the type of habitat they use for feeding and/or breeding. Some species such as the red-eyed vireo, white-breasted nuthatch, and prothonotary warbler are insectivores (feed on insects and other invertebrates), while others such as the tufted titmouse, blue jay, and American cardinal are primarily frugivores (feed on fruits and seeds). However, no species strictly adheres to this diet. Frugivores sometimes feed upon insects and other invertebrates while insectivores can make the switch to fruits and seeds when their preferred food is in limited supply. The usual habitat for many of the species listed above, including most of the warblers, is a closed canopy forest, but some species such as the American cardinal and the blue jay are much more likely to found in open forests and forest edges.

In the Appalachians and Ozarks, birds and small mammals make up the diet of various birds of prey (called raptors). The most familiar and widespread member of this group is the red-tailed hawk (*Buteo jamaicensis*), a large bird with a wingspan that can reach a meter. Two of the other more distinctive but smaller raptors are the American kestrel (*Falco sparverius*) and peregrine falcon (*F. peregrinus*). All of these birds are commonly found in open areas and forest edges. Raptors tend to be large birds, and the very largest example found in the two regions is the turkey vulture, which has a wingspan that exceed 1.8 m. However, the turkey vulture would be considered as a raptor only in the broad sense, since it feeds on carrion.

The owls represent a group of nocturnal raptors. A number of species are found in both the Appalachians and Ozarks. Two of the most common are the eastern screech owl (*Megascops asio*) and the barred owl (*Strix varia*). The eastern screech owl is a small (46–60 cm wingspan), stout bird easily recognized from the short "ear" tufts on the head and its eerie, screeching call (Fig. 7.7). Eastern screech owls are commonly gray in color but can be bright reddish-brown. Like most owls, the eastern screech owl is strictly nocturnal, spending the daylight hours in a natural cavity in a large tree, in dense foliage, old woodpecker holes, and various semi-enclosed spaces in man-made structures. The bird is probably the most common avian predator in the Appalachians and Ozarks and feeds mostly on small mammals but also preys on other animals ranging from bats to large insects. It is a permanent resident of the two regions. However, it is so secretive that most people are unaware that the bird is present until they hear its call. The barred owl is larger (96–125 cm wingspan) and can be recognized by the horizontal bars on its chest (the basis for the common name). Barred owls feed mostly upon small mammals, but sometimes preys upon other things, including birds and small reptile. The two owls differ in the types of habitats in which they occur. Barred owls are inhabitants of mature forests, whereas screen owls are more likely to be found in open woodlands and are not uncommon in urban settings.

7.3 Reptiles

7.3.1 Snakes

Three different taxonomic groups of reptiles—snakes, lizards, and turtles—occur in the Appalachians and Ozarks, with most species in each group found in both regions (Green and Pauley 1987; Johnson 2000; Trauth et al. 2004). During the winter, representatives of all three of these groups typically enter a period of dormancy.

Fig. 7.7 The reddish-brown color variant of the eastern screech owl (*Megascops asio*), a bird known for its eerie, screeching call. (Photo by Matt Orsie)

Except for turtles, which are mostly aquatic, most reptiles live on land. Reproduction occurs on land, with the embryo enclosed in an egg with a shell. Reptiles breathe by means of lungs, just like the birds and mammals, and have a skin covered with scales. In contrast, amphibians usually breathe by gills in the larval stage and lungs in the adult stage, but considerable "direct" gas exchange between the bodies of some amphibians and the surrounding air or water occurs through the skin, which is typically relatively thin and moist. Amphibians were the first vertebrates to inhabit land, but their eggs are unprotected by a shell.

Snakes are the reptiles that tend to receive the most attention, since a very few examples are venomous. Most of the species found in the Appalachians and Ozarks are harmless, but this doesn't seem to matter for most people. Humans seem to have evolved an innate tendency to fear snakes, and the usual reaction upon seeing one is to avoid it. This is clearly the proper thing to do for venomous snakes, which are not especially common, especially at the very highest elevations in the Appalachians.

There are two common venomous species of snakes in the Appalachians and Ozarks—the timber rattlesnake (*Crotalus horridus*) and the copperhead (*Agkistrodon contortrix*). The venomous northern cottonmouth (*Agkistrodon piscivorus)* also occurs in the Ozarks. These are pit vipers, a group that includes many of the more dangerously poisonous snakes in the world. Their venom is classified as hemotoxic since it affects blood (rather than the nervous system). The former is the larger of the two common venomous snakes, with an average length of 90–150 cm in adults. The copperhead has an average length of 60–90 cm. Both feed on small animals such as deer mice, birds, lizards, salamanders, and large insects. Northern copperheads have a head that is unmarked but characteristically copper-colored (hence the common name), while the body has reddish-brown hourglass-shaped cross bands that taper at midline on the back. This snake is ovoviviparous, and bears its young "alive." They are most likely to occur in dry, rocky habitats, at forest edges, and in various types of disturbed areas. The western diamondback rattlesnake (*Crotalus atrox*), the largest of the North American pit vipers, can be found in the Ozarks of Arkansas.

Copperheads are nocturnal in the summer but may be active during the day in spring and fall. Although poisonous, their bites are rarely fatal. Nevertheless, they are probably responsible for the most bites by poisonous snakes in the both the Appalachians and Ozarks as well as the entire United States. Northern copperheads do not need to be coiled in order to strike, and this should be taken into consideration if one is encountered.

The timber rattlesnake is both dangerous and impressive, and anyone having an encounter with one in nature is likely to remember it for a very long time. Although timber rattlesnakes seem to prefer rock outcrops, they can be found nearly anywhere, including the interior of a forest. This snake is nocturnal in summer but may be diurnal in spring and fall. Rattlesnakes overwinter in rocky dens with other rattlesnakes, copperheads, and/or rat snakes. Like the northern copperhead, it is not necessary for them to be coiled to strike, and they don't necessarily "rattle" as a warning before striking, as some people think. These are the most poisonous snakes in the in the entire United States.

The most common nonvenomous snakes are the ringneck snake (*Diadophis punctatus*), black rat snake (*Pantherophis obsoletus*), worm snake (*Carphophis vermis* in the Ozarks and *C. amoenus* in the Appalachians), eastern garter snake (*Thamnophis sirtalis*), and the black racer (*Coluber constrictor*). Worm and ringneck snakes are small snakes (typically 25–38 cm long) that mostly burrow under moist soil or leaf litter. Because they are infrequently observed, among the less familiar snakes in the two regions are the worm snakes. These small slender snakes superficially resemble earthworms in habitat, behavior, and general appearance. They spend most of the day beneath rocks, logs, or other cover, feeding upon soft-bodied insects and earthworms. If it is very dry and hot during the summer, they may burrow into the soil and enter a state of dormancy. Most people have never seen a worm snake and thus are unsure of what to make of it when an example is exposed when a log is turned over. These small snakes are most common in moist forests at lower elevations in the Appalachians and Ozarks.

The black racer and black rat snakes are large snakes, with adults reaching 75–180 cm long, while the eastern garter snake (Fig. 7.8) is typically half that size (45–60 cm). The latter is exceedingly common and thus is the snake most likely to be encountered in both the Appalachians and Ozarks. Because it tolerates cold temperatures, the garter snake is the earliest species to be active in the spring. Like all snakes, garter snakes are carnivores, but they prey more on invertebrates (e.g., snails, slugs, earthworms, and insects) than is the case for larger snakes. The black racer and black rat snake feed mostly on small animals, including mice, frogs, salamanders, and birds. They are somewhat arboreal. This allows them to reach the nests of birds in trees, where they can feed on eggs and baby birds. The black rat snake is the most common predator of eggs in a nest. As noted above, this snake may reach a length of 180 cm or even more, although most of the specimens one is likely to encounter are only 120–150 cm in length. The upper body is glossy black, although some traces of white and orange are visible between the scales. The underside is white but usually mottled with gray or black blotches. Juvenile black rat

Fig. 7.8 The eastern garter snake (*Thamnophis sirtalis*) is probably the most common snake in both the Appalachians and Ozarks. (Photo by Charles Garrett)

snakes have a very different color pattern consisting of a series of brown blotches on an otherwise uniformly gray background. Anyone unaware of this fact is likely to think that they belong to an entirely different species.

7.3.2 *Lizards*

Two other groups of reptiles—lizards and turtles—are not especially well represented in the Appalachians but are more common in the Ozarks. The only examples likely to be encountered with any frequency in both regions are the eastern fence lizard (*Sceloporus undulatus*) and the five-lined skink (*Plestiodon fascistus*). The former reaches a maximum length of about 17 cm and can be recognized by their gray, brown or tan body that has a series of wavy crossbands on the back. During the breeding season, males develop bright blue patches on their underside, and this accounts for the common name "blue belly" that is sometimes given to this species. Fence lizards are capable of moving rather quickly, and catching one with the bare hand is exceedingly difficult. Fence lizards are most common in dry, open forests at lower elevations in the region, but they also occur in and around houses and barns. Although often found on the ground (or a stump or log), they usually make a dash for the nearest tree when approached. Fence lizards feed upon insects and other small invertebrates. The eastern fence lizard can climb trees with ease, but the fine-lined skink stays on the ground.

The five-lined skink is slightly longer (length up to 20 cm) than the eastern fence lizard and also is very different in color (Fig. 7.9). The body is brown to black and is marked with five longitudinal yellow stripes (two on each side and one down the center of the back). Young skinks have a bright blue tail, and in adult males the head is red. One common name that has been used for this lizard is "blue-tailed scorpion" (which is sometimes shortened to "scorpion"), which is based on the totally erroneous belief that it has a venomous sting. In fact, five-lined skinks are completely harmless. Like the southern fence lizard, the diet of the five-lined skink consists mostly of insects and other invertebrates. The usual habitat for this species is a moist forest, especially a disturbed example with abundant logs and stumps. Because individuals seek cover in protected places such beneath a rock or piece of loose bark on a fallen log, they are sometimes encountered rather unexpectedly (which can be a startling experience) when the rock or log is moved. Both the five-lined skink and the fence lizard exhibit what is referred to as tail autonomy, which means that the animal can "shed" its tail if the latter is grabbed by a predator. The detached tail will continue to wiggle, distracting the predator's attention from the lizard, which is able to escape. The lizard then regenerates the tail over a period of several weeks.

The eastern collared lizard (*Crotaphytus collaris*) occurs only in the Ozarks, where it is restricted to open, rocky habitats (Fig. 7.10). This is a large lizard, with a total body length of 20–30 cm, including a very long tail relative to the length of the body. The body is brightly colored and there is a distinct black collar on the neck. This animal is very active and can jump from rock to rock in the habitats

Fig. 7.9 The distinctive five-lined skink (*Plestiodon fascistus*) is found in both the Appalachians and Ozarks. (Photo by Kim Fleming)

Fig. 7.10 The eastern collared lizard collared lizard (*Crotaphytus collaris*) occurs in portions of the Ozarks. (Photo by Eric Isley)

where it occurs. The diet consists mostly of large insects, but it is capable of feeding on small vertebrates. The eastern collared lizard is associated with igneous glades on the Springfield-Salem Plateau and in the St. Francois Mountains of Missouri.

7.3.3 Turtles

Excluding aquatic forms that would not normally occur in forested habitats, there is only one turtle that commonly occurs in the Appalachians and Ozarks. This is the eastern box turtle (*Terepene carolina*), which can be encountered in just about any habitat. Easily recognized from its high-domed, rounded carapace, box turtles are about 12–15 cm long and weigh no more than about 0.5 kg. The box turtle is fairly well camouflaged in nature by virtue of the color pattern found on the dome-shaped shell (more properly known as a carapace) that covers all but the lower surface of its body. The pattern is rather variable but consists of yellow spots and bars on a darker (brown or almost black) background. The lower surface of the body of a box turtle is protected by a second, hinged shell (known as the plastron). When disturbed, the turtle is able to withdraw its head and limbs into the protected space provided by the carapace and plastron, which affords a considerable degree of protection from most potential predators. Box turtles are omnivorous, consuming anything from invertebrates and small vertebrates (including fish, amphibians, and other reptiles) to fruits, berries, and even mushrooms. They are extremely long-lived (commonly living 50 years or even more), slow to mature, and have relatively few offspring each year. Because they are slow moving, they are particularly susceptible to being killed on highways.

7.4 Amphibians

The most abundant vertebrates in the forests of the Appalachians and Ozarks are amphibians, with salamanders the most widespread and numerous representatives of this group. In fact, the diversity of salamanders in the Appalachians is higher than for any region of comparable size anywhere in the world. Some species of salamanders occur in both regions. These include the eastern newt (*Notophthalmus viridescens*), the spotted salamander (*Ambystoma maculatum*), the marbled salamander (*A. opacum*), and the four-toed salamander (*Hemidactylium scutatum*). However, others are restricted to one region or the other. For example, the Ozark zigzag salamander (*Plethodon angusticlavius*), Oklahoma salamander (*Eurycea tyunerensis*), many-ribbed salamander (*Eucycea meltiplicata*), and western slimy salamander (*Plethodon albagula*) occur only in the Ozarks. The two most common species of salamanders in Ozarks are the Ozark zigzag salamander and spotted salamander. In contrast, the red-backed salamander (*Plethodon cinereus*) is the most common salamander in many areas of the Appalachians (Fig. 7.11). This species, along with the Cheat Mountain salamander (*P. nettingi*), Ridge and Valley salamander (*P. hoffmani*), Wehrle's salamander (*P. wehrlei*), teyanalee salamander (*P. teyanalee*), Peaks of Otter salamander (*P. hubrichti*), Shenandoah salamander (*P. shenandoah*), and mountain dusky salamander (*Desmognathus ochropheus*), is restricted to the Appalachians. Populations of some of these species, such as the Cheat Mountain salamander and the Dixie Caverns salamander (*P. dixi*), are limited to surprisingly small areas within the region. However, this is not the case for the red-backed salamander, the most widespread and abundant salamander in the Appalachians. There is no question that the Appalachians are a center of diversity for *Plethodon* salamanders, and the total number of species present is not yet known. Several species new to science (e.g., the yellow-spotted woodland salamander [*P. pauleyi*]) have only recently been described.

Members of the genus (*Plethodon*) to which several of the salamander mentioned above belong are often referred to as the "lungless" salamanders, and they actually do lack lungs. Respiration occurs almost entirely through the skin, which

Fig. 7.11 The red-backed salamander (*Plethodon cinereus*) is exceedingly common in some areas of the Appalachians. (Photo by Paul Sattler)

must remain moist if the animal is to survive. Because of their abundance, the lungless salamanders are ecologically important to the forests in which they occur in two distinct ways. First, they are major predators of insects and other invertebrates associated with the dead leaves and other plant debris on the forest floor. Second, they serve as an important source of food for larger carnivorous vertebrates, including such things as skunks, raccoons, and snakes. Some studies have shown that these salamanders should be regarded as "key" species in those forests where they characteristically occur, since a low level of abundance of these animals often reflects an "unhealthy" condition of the forest ecosystem as a whole. However, low densities in the Ozarks likely reflect relatively xeric conditions.

Most species of terrestrial salamanders, the only (with the one exception noted below) salamanders considered in this chapter, are small, usually dark colored, have a body length of no more than about 10 cm, and weigh only 1–10 g. Their usual habitat is the forest floor, where they can be found under fallen logs, other pieces of woody debris, and within thick layers of moist decaying leaves. These "woodland" salamanders feed on insects and other invertebrates while also representing an important item in the diet of small carnivores.

The focus of this book is on terrestrial habitats of the Appalachians and Ozarks, although a few special wetland habitats such as bogs and fens also are considered. However, the largest salamander in North America occurs in both regions and warrants special mention. The eastern hellbender (*Cryptobranchus alleganiensis* subsp. *alleganiensis*) is very large salamander (body length up to 70 cm and a weight of as much as 2.5 kg) found in rocky, rapidly flowing streams. This is one of the largest and most distinctive salamanders in the world. Like other aquatic organisms, the hellbender requires a stream and surrounding watershed relatively free from disturbance, primarily disturbance occurring as a result of some type of human activity. The hellbender thrives in watersheds within mature forests. In this way, they are inevitably linked to forest ecosystems although they do not directly inhabit forests. The Ozark hellbender (*Cryptobranchus alleganiensis* subsp. *bishop*), listed as endangered throughout its range, is restricted to cool, clear Ozark streams in southern Missouri and northern Arkansas. This subspecies has a much more limited distribution, but interestingly enough, Missouri does have both the eastern hellbender and the Ozark hellbender. Both species are threatened by stream degradation, pollution, predation, and disease.

Salamanders, like other amphibians, usually require water for reproduction. The members of the genus *Plethodon* represent the most noteworthy exceptions. As such, they tend to be associated with water in one form or another. The water can be a small stream, a pond, or low areas (vernal pools) where water accumulates during spring when salamanders reproduce. Some species remain close to water during their entire lives while others can be found some distance away. The same situation is true for frogs, the other group of amphibians that occurs in the forests of the Appalachians and Ozarks.

There are several frogs commonly found in the Appalachians and/or Ozarks. The American toad (*Anaxyrus americanus*), green frog (*Lithobates clamitans*), gray tree frog (*Dryophytes versicolor*), and the northern spring peeper (*Psudoacris crucifer*) are found in both regions. The American toad is a relatively large amphibian, weighing 15–35 g and easily recognized by its brown to gray colored heavily warted skin. The northern spring peeper is much smaller, weighing only about a gram. This tiny frog is brown, green, or gray in color and has a distinct X-shaped marking on its back. Spring peepers are good climbers but spend much of their time in the litter layer on the forest floor. They are usually heard a lot more often than they are seen, because in the spring they produce a "chirping" noise." It is for this reason that they are often referred to as the "chorus" frog. The mountain chorus frog (*Pseudoacria brachyphona*), another member of this group of frogs, occurs only the Appalachians.

The vertebrates discussed thus far are vastly outnumbered by the invertebrates, many of which are very small and either difficult to observe in nature or usually overlooked. The invertebrates include a huge number of insects and non-insect arthropods such as spiders and mites. Some insects such as the larger butterflies, moths, dragonflies, grasshoppers, and beetles are fairly conspicuous and thus relatively well-known, and certain insects are of considerable economic and ecological importance to the forests of the Appalachians and Ozarks. Major groups of invertebrates found in the two regions are the subject of the next chapter.

References

Green NB, Pauley TK (1987) Amphibians and reptiles in West Virginia. University of Pittsburgh Press, Pittsburgh, PA

Guilday JE (1982) Appalachia 11,000–12,000 years ago: a biological review. Archaeol East N Am 10:22–26

Hall GA (1983) West Virginia birds: distribution and ecology, Carnegie Museum of Natural History Special Publication No. 7. Carnegie Museum of Natural History, Pittsburgh, PA

Hall GA (1984a) Population decline of Neotropical migrants in an Appalachian forest. Am Birds 38:14–18

Hall GA (1984b) A long-term bird population study in an Appalachian forest. Wilson Bull 96:228–240
Healey W, Brooks RT (1988) Small mammal abundance in northern hardwood stands in West Virginia. J Wildl Manag 52:491–496
James DA, Neal JC (1986) Arkansas birds: their distribution and abundance. University of Arkansas Press, Fayetteville
Johnson TR (2000) The amphibians and reptiles of Missouri. Missouri Department of Conservation, Jefferson City
Linzey DW (2021) The mammals of Virginia. The McDonald and Woodward Publishing Company, Newark, OH
Lituma CM, Cox JJ, Spear SF, Edwards JW, De La Cruz JL, Muller LI, Ford WM (2021) Terrestrial wildlife in the post-mined Appalachian landscape: status and opportunities. In: Appalachia's coal-mined landscapes: resources and communities in a new energy era. Springer Nature, Cham, pp 135–166
Miller BF, Campbell TA, Benjamin R. Laseter BR, Ford WM, Miller KV (2009) White-tailed deer herbivory and timber harvesting rates: implications for regeneration success. For Ecol Manag 258:1067–1072
Nanavati WP, Grimm EC (2020) Humans, fire, and ecology in the southern Missouri Ozarks, USA. The Holocene 30:125–135
Neff JA (1923) Some birds of the Ozark region. Wilson Bull 35:202–215
Sasse DB (2022) Early records suggest a broader historic range of the Ozark big-eared bat (*Corynorhinus townsendii ingens*). West N Am Nat 82:155–158
Sealander JA, Heidt GA (1990) Arkansas mammals: their natural history, classification, and distribution. University of Arkansas Press, Fayetteville
Smith RL (1993) Wildlife of the upland forests. In: Stephenson SL (ed) Upland forests of West Virginia. McClain Printing Company, Parsons, WV, pp 211–229
Smith KG, Clark JD (1994) Black bears in Arkansas: characteristics of a successful translocation. J Mammal 75:309–320
Stephenson SL (2010) The kingdom fungi: the biology of mushrooms, molds, and lichens. Timber Press, Portland, OR
Thorne ED, Fies ML, Ford WM (2024) Home range and resource use by eastern spotted skunks in Virginia. JSEAFWA 11:206–214
Trauth SE, Robison HW, Plummer MV (2004) The amphibians and reptiles of Arkansas. University of Arkansas Press, Fayetteville
Wentworth JM, Johnson AS, Hale PE (1990) Influence of acorn use on nutritional status and reproduction of deer in the Southern Appalachians. JSEAFWA 44:142–154

Chapter 8
Insects and Other Invertebrates

Abstract The most biodiverse groups of organisms in the forests of the Appalachians and Ozarks are fungi (especially microfungi) and invertebrate animals. The latter fact is not surprising, since more than 90% of all animals are invertebrates. Insects represent the single most diverse group of invertebrates in the forests of the Appalachians and Ozarks. Among the more ecologically important insects are pollinators (e.g., primarily butterflies, moths, bees, wasps, flies, and beetles), but insects also have many other ecological roles, including herbivory or as predators or parasites on other insects. These sometimes can play a major role in regulating outbreaks of pest insects. Insects themselves provide food for various groups of animals, including both invertebrates and vertebrates. Common non-insect invertebrates of the Appalachians and Ozarks include other arthropods such as arachnids (spiders and mites), centipedes, and millipedes as well as representatives of other phyla. The latter include earthworms and mollusks (snails and slugs). These groups and the ecological roles they play are discussed in this chapter.

Keywords ATBI · Non-insect invertebrates · Ecological roles · Orders of insects · Arachnids · Pollinators · Predators · Mollusks

8.1 Introduction

In a forest ecosystem, trees are the dominant organisms and represent the greatest biomass. However, trees are among the least biodiverse groups of organisms present. The greatest biodiversity exists for the fungi (especially microfungi) and invertebrate animals. More than 90% of all described species of animals are invertebrates (i.e., lack an internal skeletal system), and members of this large and diverse

---

S. L. Stephenson, R.-M. Muzika, *Forests of the Southern Appalachian Highlands and Interior Highlands of the United States*, Plant and Vegetation 20,
https://doi.org/10.1007/978-3-032-01155-8_8

assemblage are well represented in the Appalachians and Ozarks. The fungi are considered in Chap. 9; the focus of this chapter is on invertebrates. The most conspicuous invertebrates in forests are insects. Some of these are large and colorful enough to be easily noticed, but many others are readily overlooked or even unknown to most people. In a practical sense, invertebrates (irrespective of the taxonomic group to which they belong) fall into two categories—microinvertebrates and macroinvertebrates. Microinvertebrates are too small to be seen with the naked eye (usually with a body length of less than 1 mm). In contrast, macroinvertebrate can be seen with the naked eye, although they may not be obvious when viewed against the substrate upon which they occur.

Invertebrates are generally smaller than vertebrates, but this isn't always the case. For example, the eastern Hercules beetle (*Dynastes tityus*), the largest beetle found in the Appalachians and Ozarks, can weigh as much as 37 g and be 60 mm long. The American pygmy shrew (*Sorex hoyi*), a rare animal in forests at higher elevations in the Appalachians but absent from the Ozarks, weighs only about 2 g and is no more than 50 mm long. Although they lack the body mass of the eastern Hercules beetle, some butterflies, moths, and dragonflies that occur in the two regions have a wingspan of 12 cm or more. The largest dragonfly found in the two regions is the giant darner (*Anax walsinghami*), with a wingspan of up to 12 cm and a body length of 10 cm. The largest butterfly is the eastern *giant swallowtail* (*Papilio cresphontes*), which can have a wingspan up to 19 cm. The largest moth, the cecropia moth (*Hyalophora cecropia*), is almost as large and can have a wingspan of 18 cm (Fig. 8.1). This moth is certainly one of the most beautiful insects in the world.

Some terrestrial non-insect arthropods also are of considerable size. For example, the American giant millipede (*Narceus americanus*) can reach a length of 10 cm (Fig. 8.2). Other groups of non-insect invertebrates found in the Appalachians and Ozarks are even larger if both aquatic and terrestrial forms are considered. Earthworms can exceed 20 cm in length, while crayfish and larger freshwater mussels have a body mass that exceeds that of any insect.

Fig. 8.1 The cecropia moth (*Hyalophora cecropia)* is certainly one of the most beautiful insects found in the Appalachians and Ozarks if not the entire world. (Photo by Timothy Hoeflich)

Fig. 8.2 The American giant millipede (*Narceus americanus*) is one of the largest non-insect terrestrial invertebrates and can reach a length of 10 cm. (Photo by Katja Schulz)

8.2 All Taxa Biodiversity Inventory

In December of 1997 a group of scientists, educators, and administrators met in Gatlinburg, Tennessee to discuss the idea of carrying out an all taxa biodiversity inventory (ATBI) of the Great Smoky Mountains National Park (Sharkey 2001). In brief, this would be an effort to document all of the species of living organisms in the entire park, considering both aquatic and terrestrial habitats. The ATBI is still ongoing 40 years later, and the current total number of species known from the park (as of December 2024) is 22,143. The Great Smoky Mountains National Park falls within the very heart of the Appalachians as defined in this book. No comparable ATBI has ever been carried out anywhere else in the world. The Great Smoky Mountains encompass only a portion of the Appalachians, and there are no comparable data for any region of the Ozarks, although Natural Heritage programs in Arkansas and Missouri have provided a considerable body of data on the occurrences of species in all groups of organisms. The data from the ATBI provide a valuable insight into the relative numbers of species for various groups of invertebrates in the temperate forests of the eastern United States. It should be noted that these numbers do not distinguish between terrestrial and aquatic species. Nevertheless, they are still useful for comparative purposes.

8.3 Insects

Insects (class Insecta) are the most diverse taxonomic group of terrestrial animals, with about one million different species having been formally described. However, it is generally accepted that the actual number of species globally is much higher. Although many insects (e.g., butterflies, moths, dragonflies, and beetles) would be considered as macroinvertebrates, there also are many (e.g., springtails and other soil-inhabiting invertebrates) that are clearly microinvertebrates. This same situation (i.e., a particular taxonomic group encompassing both microinvertebrates and macroinvertebrates) applies to many of the other taxonomic groups discussed in this chapter.

As is generally known, the major distinguishing feature of an insect is that it has six legs. Most adult insects are capable of flight, and this—coupled with their sheer abundance—makes them more conspicuous than any other group of invertebrates found in the forests of the Appalachians and Ozarks. Although most insects are terrestrial, numerous aquatic forms also exist, including some that are aquatic when immature but develop into terrestrial adults. Since the focus of this book is on terrestrial organisms, only a few of the latter will be considered herein. There are approximately 30 different and (usually) readily distinguishable taxonomic groups (orders) of insects (Wheeler et al. 2001; Johnson and Triplehorn 2004). Only a few of these are ecologically important enough to be considered in this book.

8.3.1 Beetles

It has been estimated that approximately 40% of all insects are beetles (order Coleoptera). In the Great Smoky Mountains National Park, 2709 species of beetles have been recorded, the highest total for any group of insects (Discover Life in America 2024). The abundance of members of this order is readily apparent in the Appalachians and Ozarks, where there are at least some beetles associated with every imaginable type of ecological habitat. Removing a loose piece of dead bark from a decaying log, turning over the log, sorting through the dead leaves on the forest floor, checking old fruiting bodies of fungi, or examining the flowers and leaves of living plants invariably yields several types of beetles.

Beetles have numerous ecological roles. Bark beetles and wood-boring beetles help to break down dead plant material, carrion beetles feed upon the dead bodies of animals, and leaf beetles feed upon living plants. In addition, certain small beetles such as fungus beetles feed upon fungi, and numerous species, both large and small, are predators of other insects and other invertebrates. Although beetles do not immediately come to mind when one thinks about the role of insects as pollinators, flowers pollinated by beetles are not uncommon. Such flowers are large, produce abundant pollen, and have a strong odor. Beetles are rather inefficient ("sloppy") pollinators when compared to the other groups of insects that perform the same service to plants, but it would be a mistake to discount their contribution. Goatsbeard (*Aruncus dioicus*), which commonly grows on moist roadside banks and along forest edges in early summer, is especially noteworthy for attracting a large number of beetles to its flowers, and the same was once true of the flowering catkins of American chestnut (*Castanea dendata*) in the Appalachians.

Beetles that feed on phloem or xylem can be extremely damaging to trees. One of the most notable examples in the southeastern United States, including portions the Ozarks of Arkansas, is the southern pine beetle (*Dendroctonus frontalis*), which has caused mortality of pines and substantial economic damage over the past several decades. Other beetles, such as turpentine beetle (*Dendrotonus terebrans*), are often a secondary pest when trees have been stressed. Ambrosia beetles are notable in that they cultivate their own food (a fungus) rather than consuming phloem directly. Often more of a wood degrading than mortality-causing group of beetles, some ambrosia beetles have been found to cause substantial mortality, although it is actually the fungus they cultivate that is the mortality agent. A notable example is laurel wilt disease, caused by the fungus *Raffaelea lauricola*, which is "grown and dispersed" by the redbay ambrosia beetle (*Xyleborus glabratus*). Both are likely non-native to North American but can be devastating and cause mortality of members of the plant family Lauraceae such as redbay (*Persea borbonia*) and sassafras (*Sassafras albidum*).

As a group, the beetles include a large number of predatory forms, including ground beetles (family Carabidae) and rove beetles (family Staphylinidae). The families to which these beetles belong are among the largest families of beetles. Both ground beetles and rove beetles are very active insects, but since many of them are nocturnal, this is not readily apparent. During the summer months, anyone who takes the time to go out at night with a flashlight is likely to be surprised at the numbers of beetles scurrying about on a forest floor. Representatives of the two groups of beetles are easy to distinguish. Rove beetles are generally smaller (usually less than 20 mm in length), have slender bodies, and are black or brown. Their distinguishing feature is a wing cover that leaves half or more of the abdomen exposed. Ground beetles can be larger (often exceeding 20 mm in length), have wider bodies, and are often shiny black. Most predatory beetles are generalists with respect to the prey they consume, but some target particular groups. For example, both larvae and adults of the ladybug beetles (*Coccinella* spp.) feed largely on aphids (family Aphididae). Ground beetles (*Calosoma* spp.) are often referred to as "caterpillar hunters" because some species feed almost exclusively on caterpillars.

8.3.2 Flies

The general term "fly" is applied to a number of insects including butterflies, dragonflies, and stoneflies, but the only true flies are members of the order Diptera. Flies are the second largest group of insects (a total of 2356 species) recorded in the ATBI. The word "Diptera" translates as "two wings," and the defining characteristic of flies is that they appear to have only two wings. (The hind pair of wings is greatly reduced in size, appearing as tiny, knoblike structures that help to stabilize flight.) As a group, flies are mostly small, soft-bodied insects, but as anyone who has ever attempted to swat a persistently annoying fly knows, many are strong fliers capable of rapid flight and amazing feats of aerial acrobatics. Moreover, most flies have feet equipped with claws and pads that enable them to land easily and walk about on the lower surface of leaves and other structures.

Their flying prowess undoubtedly contributes to the success of flies, but the real key to their success is the extraordinary variety in life histories exhibited within the group and the wide range of food sources they can exploit. Flies undergo complete metamorphosis, and the larvae are generally referred to as maggots. Some are so wormlike that they could be mistaken for annelids, the phylum to which earthworms belong.

A large proportion of flies have larvae that live in aquatic habitats. Adult flies have mouthparts that are designed to suck or pierce; as a result, they can obtain their food only in liquid form. This can be nectar from flowers, plant and animal debris that has been liquefied through decomposition or, in some cases (e.g., mosquitoes), the blood of vertebrates. Many of the flies that visit flowers to feed upon nectar play a major role in pollination. Flowers pollinated by flies fall into two categories. The first consists of flowers that are brown or orange and characterized by a strong, unusually unpleasant odor. Examples include skunk cabbage (*Symplocarpus foetidus*), pawpaw (*Asimina triloba*), and jack-in-the-pulpit (*Arisaema triphyllum*). The second category is made up of flowers that are purple, violet, or blue and lack a strong odor. These features overlap with the characteristics of bee-pollinated flowers. Because of this, it is not unusual to observe a flower being visited by both bees and flies. This is especially true if the flies are members of the family Syrphidae, the flower flies or hover flies. These brightly colored flies are important pollinators, and many species resemble bees.

The association of flies with decaying plant and (especially) animal matter is impossible to overlook in nature. The flies that consume decaying matter are best regarded as scavengers. In numerous species both larvae and adults are scavengers, but in many others only the larvae are scavengers, and the adults feed upon the nectar of flowers. Some flies are active predators and use other insects as food. This is the case for robber flies (family Asilidae), the adults of which attack a wide range of insects, including other flies, beetles, butterflies, moths, bees, and wasps.

Among the more distinctive flies are the craneflies (family Tipulidae). These are large flies (their bodies sometimes reaching a length of 2.5 cm) with extremely long legs. Their slender bodies are similar in general appearance to those of mosquitoes, and some people have mistaken craneflies for large mosquitoes. However, mosquitoes never reach the size of craneflies, and perhaps more importantly, craneflies do not bite. Few craneflies ever feed as adults, and the ones that do feed upon the nectar of flowers. These long-legged flies are frequently encountered in moist, sheltered environments in forests such as the large cavities of decaying hollow logs.

8.3.3 Butterflies and Moths

A total of 1966 species of butterflies and moths have been recorded thus far in the ATBI, making the order Lepidoptera the third largest group of insects. Butterflies, especially the larger forms, are the most conspicuous insects and have been the subject of numerous field guides (e.g., Spencer 2006). In general, moths are less noticeable than butterflies but are much more numerous (worldwide, there are about ten times as many species of moths as butterflies). Moths are usually active at night and are attracted to light, while butterflies are usually active during the day. However, this generalization does not always hold true since some moths are active during the day. Such is the case for the bumblebee moth (*Hemaris* spp.) which resembles a bumblebee (*Bombus impatiens*) in both appearance and behavior (Fig. 8.3). This species is most likely to be encountered in old fields and forest openings throughout the Appalachians and Ozarks. The bumblebee moth produces the same sort of buzzing sound that one associates with bees, has an abdomen with yellow and black segments much like that of a bumblebee, and wings that are mostly transparent (although the veins and margin are dark brown). The bumblebee moth beats its wings very rapidly to fly, another bee-like feature, and it hovers over a flower to feed. Upon seeing one for the first time, few people would think of this insect as a moth. It is an important pollinator.

Both butterflies and moths undergo complete metamorphosis, as do almost 90% of all insects. This means that an individual goes through four stages—egg, larva, pupa, and adult—in its life cycle. The larvae of butterflies and moths are referred to as caterpillars. Almost all of these feed upon plant tissues, sometimes with disastrous results for the host plant. Some species are major agricultural pests, while others are destructive to forest trees as will be noted below. Many caterpillars are selective about what they will eat, and some feed on only one species of plant in their lifetime. If the host plant is uncommon or has a limited distribution, the species of butterfly or moth faces a severe constraint when it comes to finding a suitable host plant on which to deposit its eggs. This is the case for several rare species found in high-elevation wetlands in the Appalachians. Prominent examples include the bog copper (*Lycaena epixanthe*), a butterfly associated with cranberry (*Vaccinium* spp.), and Harris' checkerspot (*Chlosyne harrisii*), a butterfly whose host plant is the flat-topped white aster (*Aster umbellatus*).

Adult butterflies and moths are harmless, but the same is not the case for the caterpillars from which they develop, some of which can have a considerable negative economic impact on forests. The spongy moth (*Lymantria dispar*), a species introduced to North American from Europe, is undoubtedly the best-known

Fig. 8.3 The bumblebee moth (right side of image) resembles a bumblebee (left side of image) in both appearance and behavior. (Photo by Dan Laughman)

defoliator of hardwood trees, but there are numerous other examples. Among the more common are native defoliators such as the fall cankerworm (*Alsophila pometaria*), linden looper (*Erannis tiliaria*), hickory tussock moth (*Lophocampa caryae*), and half-wing geometer (*Phigalia titea*). Each has the potential to completely defoliate a heavily infested tree. The most easily recognized member of this group of defoliators is the eastern tent caterpillar (*Malacosoma americanum*), which constructs conspicuous silken, tent-like communal nests in the branches of black cherry (*Prunus serotina*) and related trees such as apple (*Malus* spp.).

Adult butterflies and moths usually feed on the nectar of flowers. As a result of their feeding activities, they have considerable ecological importance as pollinators. Flowers pollinated by butterflies tend to be large and brightly colored, which makes the association of butterfly and flower even more apparent. Flowers most likely to be visited by butterflies often grow in clusters, produce abundant nectar, and are arranged in such a way that the insect has a place to land. Butterfly-pollinated flowers are most abundant at forest edges and in open areas in late summer and fall; these include many members of the family Asteraceae, which have flowers with all the features described above. Prominent examples include sunflowers (*Helianthus* spp.), true asters (*Aster* spp.), goldenrods (*Solidago* spp.), and common ironweed (*Veronia fasciculata*).

Among the larger and thus more noticeable butterflies visiting these flowers are swallowtails (family Papilionidae), a group of colorful butterflies named for the long "tails" on their hindwings that resemble the long, pointed tails of swallows. One of the most familiar is the eastern tiger swallowtail (*Palilio glaucus*), which is easily recognized from the distinctive yellow-and-black-striped markings on its wings and body (Fig. 8.4). The eastern tiger swallowtail was the first insect from North America to be formally described by science.

Moth-pollinated flowers have many of the same features as butterfly-pollinated flowers, but most are white (since colors cannot be discerned in the dark) and

Fig. 8.4 The eastern tiger swallowtail (*Palilio glaucus*) is easily recognized from the distinctive yellow-and-black-striped markings on its wings and body. (Photo by Steven L. Stephenson)

produce a strong, sweet odor that can be detected easily by potential visitors. Two examples are the common evening primrose (*Oenothera biennis*) and some species of morning glory (*Ipomea* spp.). Since the association of most moths with flowers is generally apparent only at night, it goes largely unnoticed, but there is no question of its ecological importance. Among the more commonly encountered night-pollinating moths are members of the family Noctuidae (noctuids or owlet moths), the largest single group within the order Lepidoptera. Noctuids are stout-bodied, usually dull-colored moths; they make up the majority of the large moths attracted to lights at night.

8.3.4 Bees, Wasps, and Ants

Bees, wasps, and ants belong to the insect order Hymenoptera. A total of 1579 species in these three groups have been recorded in the ATBI. This makes the Hymenoptera the fourth largest group of insects. More than 12,000 species of ants occur throughout the world, and ants are commonly encountered in all types of habitats (Lessard et al. 2007). It is impossible to know exactly how many species occur in either the Appalachians or Ozarks as delimited in this book. However, more than 100 species are known from all of Arkansas and about 150 species from all of Missouri. One hundred and eight species have been recorded from just the Great Smoky Mountains National Park (Discover Life in America 2024).

Ants are critically important to many wildflowers, including examples such as Dutchman's breeches (*Dicentra cucullaria*), bloodroot, violets (*Viola* spp.), spring beauty, trillium, hairy disporum (*Prosartes lanuginosa*), trailing arbutus (*Epigaea repens*), liverleaf (*Hepatica* spp.), wild ginger (*Asarum canadense*), and trout lily. These are only some of the plants found in the forests of the Appalachians and Ozarks in which the seeds depend upon ants for dispersal (Karnish 2024). Although quite common in nature, this type of dispersal wasn't recognized by scientists until a few decades ago. It is now estimated that at least a third of the species of forest understory plants in the eastern United States have their seeds dispersed by ants (Gaddy 1986). How does this work? First, each seed has an attached "food body" (the technical term is elaiosome) that attracts ants. The ants take the seeds back to their nests, where the food body but not the seed is eaten. The seeds end up in the nest debris, where they germinate and can produce a new plant some distance away from the plant on which they were produced. The nest debris provides an especially favorable situation for germination and early growth of the new plant to occur. This relationship between plants and ants is referred to as myrmecochory.

Most ants are often little more than a nuisance when they occur in places where they are not wanted, but one ant that has recently spread into the Appalachians and Ozarks is more than a nuisance. This is the fire ant (*Solenopsis invicta*), which is a serious pest. This species is native to South America but was introduced to North America in the 1930s. Fire ants were first reported from southern Arkansas in 1958 and soon spread northward into the Ozarks. They had reached lower elevation regions of the Appalachians by the end of the twentieth century and are now being found in areas at higher elevations. Fire ants can inflict painful stings and are capable of attacking and killing small animals. The presence of their large nests can decimate local insect populations and affect the composition of the surrounding vegetation.

Although other groups of insects are involved in pollinating plants, the most important pollinators are members of the order Hymenoptera, especially bees. Because bees are a diverse group of insects, bee-pollinated flowers exhibit a wide range of sizes, shapes, and colors. Many are yellow or blue and have petals marked with lines or spots (known as nectar guides) that presumably guide a potential pollinator to the portion of the flower where pollen or nectar can be obtained. Often these nectar guides are visible only under ultraviolet light, a part of the color spectrum that bees can see but humans cannot. Nevertheless, nectar guides are clearly apparent in a number of flowers, with the spring beauty (*Clintonia* spp.) an especially noteworthy example. Bees appear to be the primary pollinators of many spring wildflowers, including such familiar examples as bloodroot (*Sanguinaria canadensis*), the various species of trillium (*Trillium* spp.), yellow trout lily (*Erythronium americanum*), and coltsfoot (*Tussilago farfara).* However, butterflies appear to be more important for a few examples such as toothworts (*Dentaria* spp.)

Some of the bees that pollinate specialize on the flowers of a particular genus of plant, and this is true for species found in the Appalachians and Ozarks (Coleman Little, pers. comm.). For example, the spring beauty miner bee (*Andrena erigeniae*) collects pollen from species of *Claytonia* (e.g., *C. virginica*) and has observed only rarely on flowers of other species. The southeastern blueberry bee (*Habropoda laboriosa*) regularly visits species of blueberry (*Vaccinium* spp.), the goldenrod cellophane bee (*Colletes solidaginis*) is largely restricted to goldenrod (*Solidago* spp.), and the passionflower bee (*Protandrena passiflorae*) is almost always associated with the yellow passion flower (*Passiflora lutea*).

The term "invasive" usually has a negative connotation when it refers to plants, animals, or fungi introduced to North America from elsewhere in the world. However, there is one invasive bee that has had a major positive effect on the ecology of much of the continent. The honeybee (*Apis mellifera*) was introduced to North America from Europe in the 1700s. Since its introduction, the honeybee has become an important pollinator and also a source of a valuable commodity—honey. Populations of honeybees have experienced a noticeable decline since the mid-1980s, which is concerning.

Because they are known to sting, wasps (at least the larger examples) are familiar insects to almost everyone. Unfortunately, few people are aware of the ecologically important role that most wasps play in nature by controlling the number of other insects. These wasps are sometimes referred to as parasitic wasps, but the term parasitoid wasp is more appropriate. A parasitoid is an organism that first feeds upon but then ultimately kills its host. The adult female parasitoid wasp deposits an egg on or inside the body of its host, typically the larva of another insect but sometimes an adult insect or non-insect arthropod such as a spider. When the egg hatches, the larval wasp feeds on the tissues of the host, gradually killing it. Until quite recently the incredible number of species of parasitoid wasps was not fully appreciated. We now know that they represent the largest group within the order Hymenoptera if not the largest group within the insects as a whole. Most species attack a restricted range of hosts, sometimes only the caterpillars of a single species of butterfly or moth. Since there are few groups of insects that are not attacked by parasitoid wasps, the number of species is presumably immense. Parasitoid wasps are mostly small, with many species with bodies only a fraction of a millimeter long. The most commonly encountered of the larger forms are members of the family Ichneumonidae, commonly referred to as ichneumons or ichneumon wasps. Ichneumons are generally black or brown, but some have yellow, orange, or red markings. The most noticeable features are the often long antennae on the head and especially the ovipositor at the end of the abdomen, often exceedingly long relative to the rest of the body. Although their long ovipositor gives these wasps a formidable appearance, they are harmless to humans. The wasps use their long ovipositor to bore into wood to reach and parasitize other insects.

8.3.5 True Bugs

The true bugs (order Hemiptera) are another large and important but usually underappreciated group of insects. A total of 922 species have been recorded by the ATBI, which is sixth highest total recorded for any group of insects. The best-known true bugs are cicadas and stink bugs (a type of shield bug), but the true bugs also include such exceedingly common examples as other types of shield bugs, aphids, assassin bugs, plant-hoppers, and leafhoppers. A defining characteristic of the order is the presence of sucking mouthparts, which form a beaklike structure. Most true bugs use their mouthparts to feed upon plants, but numerous examples, including the appropriately named assassin bugs (family Reduviidae), are predaceous on other insects. A few true bugs are hematophagous and feed upon the blood of vertebrates, including humans. The best example of the latter is the bed bug (*Cimex lectularius* and *C. hemipterus*), which is an insect associated with human dwellings and not forests.

True bugs are found in a wide variety of terrestrial habitats, although a few have become adapted to life in or on the surface of fresh water. The most abundant but lesser known true bugs are leafhoppers and plant-hoppers, which occur on almost all types of plants. These are all small insects, and few exceed 6 mm in length. One can get some indication of how common they are by simply looking down when passing through a grassy glade or meadow during the summer months. Most of the tiny things that can be observed "hopping" away as the grasses and other plants are disturbed are plant-hoppers or leafhoppers.

In many true bugs, the wings are held flat over the back, with the front wings consisting of a leathery basal portion and a membranous apical portion. The wings fit together in such a fashion that a small triangular structure called a scutellum is usually visible between the two wings. It is often possible to identify an unknown insect as a true bug by simply noting the presence of this triangle.

Fig. 8.5 Dragonflies are among of the larger and more conspicuous of insects, especially when they are flying. (Photo by Katja Schulz)

8.3.6 Dragonflies and Damselflies

Dragonflies and damselflies are large, conspicuous insects that belong to the order Odonata. Although many orders are more species rich than the 101 species recorded for these two groups from the ATBI, dragonflies and damselflies are far better known and more likely to be noticed than representatives of these other orders. Dragonflies and damselflies are usually found close to fresh water, which is where they live until they become adults. Nevertheless, some large dragonflies can be observed a considerable distance from water. Dragonflies and damselflies differ in a number of ways. Dragonflies are generally larger and more robust than damselflies and are capable of rapid and agile flight. When at rest, the wings are held out at a right angle to the body (Fig. 8.5). Damselflies tend to be smaller, more delicate looking, and have a thin abdomen. Their flight is relatively slow and weak compared to dragonflies. When at rest, the wings are held together along the length of the abdomen. Dragonflies are aggressive and effective aerial predators, feeding upon various insects such as mosquitoes, flies, butterflies, and bees. Because of their relatively slower and weaker flight, damselflies are more limited with respect to their prey. They feed primarily on small flies, including mosquitoes.

8.3.7 Springtails

The springtails (order Collembola) are a group of primitive, wingless and very small (often a length of no more than a millimeter) insects that usually go unnoticed in nature (Fig. 8.6). However, these tiny insects are incredibly common in leaf litter on the forest floor, where they feed upon fungi and decaying organic matter. Springtails frequently occur in large numbers on the gills of mushrooms, where they feed upon the spores. The major distinguishing feature of a springtail is a special structure

Fig. 8.6 Springtails are primitive, wingless, and very small (often a length of no more than 1 mm) insects that usually go unnoticed in nature. (Photo by Kim Fleming)

called a furcula on the lower side of the abdomen that acts like a spring and enables the insect to jump or spring (hence the common name) a considerable distance, sometimes more than 20 times the length of the body in some species.

The springtails represent an excellent example of what we don't know about the diversity of a forest unless we take a very close look. There are now, as a result of the ATBI, 288 species of springtails known from the Great Smoky Mountains National Park. Prior to the ATBI, 89 species had been recorded from the park. This means that the survey efforts carried out in the context of the ATBI added 138 new species of springtails. More remarkably, 61 of these were not only new to the park but also new to science! The same thing has been noted for other groups of organisms. Clearly, the forests and non-forested habitats of the Appalachians are more diverse than could have been imagined, and there is no reason to think that the total biodiversity of any group of invertebrates has been determined except for some of those that contain the very largest and most obvious species.

8.3.8 Other Insects

Representatives of many other orders of insects occur in the forests of the Appalachians and Ozarks, and couple of these should be mentioned. One for its ecological importance and the other because they are familiar to most people. About 3000 species of termites (order Isoptera) are known to science, but this group is represented by very few species in the Appalachians and Ozarks. However, individuals of these few species are both common and important to the ecology of forests in the two regions. Although sometimes referred to as "white ants," they are quite different from ants. In fact, termites are more closely related to cockroaches than to ants. Termites are rarely observed in a forest because of their cryptic life-history. They are soft-bodied and remain hidden within the galleries and tunnels of their nests for most of their lives. Termites feed upon various types of plant material, including wood, leaf litter, and soil humus, but are best known for helping to degrade wood. Termites are among the first organisms to colonize a decaying log, and through their feeding and tunneling activities, the structure of the dead wood is modified by the creation of galleries and tunnels. Once these have been established, they facilitate colonization by larger wood-feeding invertebrates. Termites are sometimes referred to as "soil engineers" because their activities translocate mineral soil to the surface, result in the construction of large pores to improve water infiltration, increase the mineral and organic carbon content of the soil, and facilitate the growth of microbes such as fungi and bacteria (Hyseni and Garrick 2019).

Praying mantids (order Mantodea) are predatory insects named for their unique forelimbs which resemble a praying position. The Carolina mantis (*Stagmomantis carolina*) is found throughout the Appalachians and Ozarks but is common only at lower elevations in the former region. They can be found worldwide. The Carolina mantis is a native species, but the two mantids most likely to be encountered are two introduced species—the Chinese mantis (*Tenodera sinensis*) and the European mantis (*Mantis religiosa*). The former was accidentally introduced into the United States in 1896, while the latter was intentionally introduced as pest control for the spongy moth. The Chinese mantis is the larger of the two species and can reach a length of 12 cm. Mantids primarily feed on other insects, but Chinese mantis also has been known to feed on small reptiles and amphibians.

8.4 Non-insect Arthropods and Other Invertebrates

8.4.1 Arachnids

The arachnids (class Arachnida) include spiders and several other groups of non-insect arthropods, the most common and widespread of which are mites, ticks, scorpions, and harvestmen. Arachnids are characterized morphologically by being flightless, having a body divided into two parts, and four pairs of legs. Ecologically, except for a few aquatic forms, all members of this class are terrestrial predators or (less commonly) parasites. The species known from the Appalachians and Ozarks range in size from very small (0.08 mm for some tiny mites) to more than a couple of centimeters for the largest spiders and the one scorpion found in the two regions. Worldwide, it has been estimated that there are at least 150,000 formally described species of arachnids, about one third of which are spiders.

Although this fact is not generally appreciated, the mites are by far the largest group of arachnids in terms of number of individuals, and they occur in both terrestrial and aquatic habitats. Since most mites are microscopic, they go largely unnoticed, but if a loose handful of leaf litter collected from the forest floor is shaken over a piece of white paper, it will invariably yield numerous tiny arthropods. Most are so small that they appear only as "specks" on the surface of the paper. A closer examination will reveal that many of the darker specks, which can be seen moving about (albeit quite slowly), are mites. Many of these are oribatid mites, the most abundant of the various types of mites that live in the humus and litter on the forest floor, sometimes numbering in the tens of thousands in a single square meter. These organisms feed mostly upon fungi and dead plant material. The ecological contribution of these tiny detritivores is rarely appreciated, but it is of immense importance to nutrient cycling in forests and other ecosystems. Not all oribatid mites are detritivores. Some are miniature carnivores and prey upon other tiny arthropods that share the same habitat.

Ticks resemble mites but are much larger. More importantly, ticks are parasites of mammals, birds, and reptiles. The most familiar ticks to most people are those found on mammals, including humans. In nature, ticks crawl up leaves and stems of grasses and other low-growing plants and wait for potential host animals. When the plant is brushed by a moving animal (or person), the tick quickly lets go of the plant and transfers to the host, where it feeds on blood. Unfortunately, ticks can carry diseases, which can be passed along to the host. The most common tick-borne disease in the Appalachians is Lyme disease; however, there are many other tick-borne illnesses with new one seeming to arise each year. In the Ozarks Lyme disease is not common but many other tick borne diseases are. The American dog tick, sometimes called the wood tick, is the most frequently encountered member of this group. Anyone who spends a reasonable amount of time hiking and camping is almost certainly familiar with it. The body of this species is reddish brown, with white or yellow markings, and is about 1.5 mm long. The American dog tick is uncommon at the very highest elevations in the Appalachians but sometimes exceedingly abundant in grassy areas and at the edges of forests at lower elevations. In the Ozarks, occurrence of this animal has no elevational constraint.

Chiggers are the larvae of a group of mites called harvest mites. The adult harvest mites lay their eggs on aerial portions of low-growing plants. Upon hatching, the larvae remain on the plant until they come into contact with some passing animal that represents a suitable host. Just like ticks, they transfer to the host. Chiggers are red but barely visible to the naked eye. They migrate over the skin of the host until they find an optimal feeding area. The feeding activities of the chigger result in the destruction of skin tissue, which causes substantial itching, as anyone who has ever picked up a chigger knows all too well!

Although mites may turn out to be the group of arachnids with the largest number of species when they have been surveyed more completely, spiders are the largest group currently known, and they exhibit an extraordinary diversity of sizes and forms. At least some examples can be found in virtually every type of terrestrial habitat. Yet, in spite of their wide range of size and forms, spiders can be recognized by a body divided into two clearly discernable regions—an anterior cephalothorax and a posterior abdomen. Attached to the lower front of the cephalothorax are jaws with fangs, which the spider uses to inject poisonous venom into its prey. The venomous nature of spiders is greatly exaggerated, and in only a very few species is spider venom poisonous enough to be dangerous to humans.

There are two species of spiders found in the Appalachians and Ozarks that are considered dangerous or "medically significant." These are the female southern black widow (*Latrodectus variolus*) and the brown recluse (*Loxosceles reclusa*). Both are venomous enough that their bites require immediate medical attention. The venom injected by the southern black widow is highly toxic (15 times stronger than the venom of a rattlesnake!). Southern black widow spiders are easily recognized from their shiny black bodies and the distinctive red hourglass-shaped mark on the abdomen. Fortunately, these spiders are not aggressive and rarely bite humans. The brown recluse is less distinctive but often can be identified from a combination of features. These are the yellowish to grayish brown body, a darker head with faint or distinct violin-shaped marking, and the relatively long legs. The markings on the head account for the two other common names (violin spider or fiddle back) given to this spider.

A few other spiders should be avoided. Although their venom is generally not dangerous for healthy individuals, it can cause swelling, severe pain, and redness. Two examples are the tiger wolf spider (*Tigrosa aspersa*) and the black and yellow garden spider (*Argiope aurantia*). Both are large spiders, reaching a length of more than 7 cm but occur in different situations. The tiger wolf spider doesn't construct a web and hunts its prey on the ground. It is black or brown and actively moves about on the forest floor. In contrast, the black and yellow garden spider constructs a large web well above the ground and is not very active. The black and yellow abdomen with white spots and brownish-red legs make this spider easy to recognize. Fortunately, few people ever suffer a bite from a spider.

Spiders feed mostly on insects and other spiders caught in their webs. All spiders produce strands of what is known as spider silk, which is very thin but incredibly strong (in proportion to its weight it is much stronger than steel). Spiders produce silk through tiny pores in structures called spinnerets, located near the tip of their abdomen, and use it to construct their webs. Although anyone who goes hiking through a forest will be familiar with the experience of encountering a web—seemingly always in one's face—a much better impression of just how common webs are can be gained by going out early in the morning during the summer when there has been a heavy dew or fog the night before, because condensation on the spider webs makes them more visible. Along grassy roadsides and in fields, spider webs appear to be everywhere. Not all spiders use webs to ensnare their prey. Some species such as wolf spiders stalk and capture their prey directly, much like the animal for which they are named.

Scorpions are not common in the Appalachians and Ozarks. Only a single species, the southern unstriped scorpion (*Vaejovis carolinianus*), is found in the Appalachians, and it is rare. The striped bark scorpion (*Centruroides vittatus*), the most widely distributed scorpion species in the United States, is the only scorpion to occur in the Ozarks. However, it is relatively common in the western drier portion of the region. The striped bark scorpion is about 6 cm long and feeds primarily on spiders and insects.

Most people have never heard of pseudoscorpions, which are common in the Appalachians and Ozarks. Pseudoscorpions have a flat, pear-shaped body and pincer-like claws that resemble those of true scorpions, but they lack a scorpion's stinger and are so small that they are usually overlooked (they are often no more than 2–8 mm in length). They are completely harmless to humans and feed upon various other small arthropods. Pseudoscorpions are commonly associated with forest floor litter and also can be found on the bark surface of a living tree, especially bark that is rough and has a partial cover of mosses and lichens.

Pseudoscorpions are widely distributed. Since they are wingless and so very small, their ability to travel would seem limited, but they are able to travel considerable distances because of their phoretic ecological relationship with other species of organisms. They attach themselves to the body or (especially) legs of insects such as flies, beetles, and warps and other arthropods, especially harvestmen. Phoretic relationships are not uncommon; some types of mites also hitchhike on other animals.

8.4.2 Millipedes and Centipedes

Both millipedes and centipedes have elongated bodies with numerous legs. Their main distinguishing feature is that centipedes have one pair of legs per body segment while millipedes have two pairs. The two groups differ in several other respects. First, millipedes tend to have cylindrical bodies and move very slowly, while centipedes often have somewhat flattened bodies and can move quickly. Second, the vast majority of millipedes are detritivores and thus harmless to humans, although some species, if handled roughly, give off an ill-smelling fluid that can discolor the skin. In contrast, centipedes are carnivores and venomous, preying upon other small invertebrates, which they capture with the aid of a pair of claw-like appendages that are actually modified legs. Larger centipedes are capable of inflicting a painful bite if handled.

8.4.3 Mollusks

Several other groups of invertebrates that might not immediately come to mind are well represented in the Appalachians and Ozarks. One of these is the phylum Mollusca (mollusks), which includes snails and slugs. Land snails are common and can be encountered in all types of habitats. These animals have a coiled shell that is the defining feature of this group of mollusks. The assemblage of land snails associated with the forests of the Appalachians and Ozarks is fairly diverse (Hubricht 1972; Oesch et al. 2013; Dourson 2015), but this is not necessarily apparent since most land snails are small and generally nocturnal. As a result, this is still an understudied group, and some of the snails in the two region probably represent species not yet known to science.

Land snails require moist conditions to survive. During the day they seek shelter beneath dead leaves on the forest floor, under loosely attached pieces of bark on decaying logs, in rock crevices, and in other protected places, emerging at night to move about (albeit slowly). The species found in the two regions vary widely in size. The largest examples have a shell about the size of a quarter, but in some of the very smallest land snails the shell is no larger than the head of a pin. Some land snails are carnivorous, but most feed on living plants or are detritivores. Land snails are an important food source for larger animals such as beetles, salamanders, turtles, small mammals, and birds. Some animals are known to eat empty snail shells for the calcium they contain.

Several species of land snails known from the Appalachians and Ozarks have a very restricted distribution. One of these is the Cheat three tooth snail (*Triodopsis platysayoides*), which is known to occur only in the Cheat River Canyon on the Appalachian Plateau of West Virginia. This snail is considered to be one of the rarest land snails in the world. The same is likely to be the case for Clench's middle-toothed land snail (*Mesodon clenchi*), which is known only from a small area of the Ozarks in Arkansas.

Slugs are similar to snails in general structure but lack an external calcareous shell and therefore cannot, like snails, withdraw their body into their shells to protect against desiccation. Instead, the slug produces a layer of mucus (slime) over the entire surface of the body. This slimy layer retains moisture, so that the slug does not dry out. Nevertheless, it is effective only if the slug stays away from exposed, dry places, especially during the day. As a result, slugs spend the day hidden away in crevices such as those found between rocks, beneath pieces of loose bark, or in decaying logs on the forest floor. Because the soft body of a slug is rather compressible and there is no shell, even a large slug can squeeze itself into a much smaller place than a snail of comparable size would be able to do. It is not unusual to encounter, often unexpectedly, a large slug in an amazingly small space. Various species of slugs feed upon dead plant matter, living plants, and the fruiting bodies of fleshy fungi. They are subject to being an item in the diet of the same types of animals that feed on snails.

8.4.4 Various Types of Worms

The term "worm" is often applied to any animal with an elongated, more or less cylindrical body and no obvious legs. Many larval insects are commonly mistaken for worms, but they develop into adults that look nothing like worms. Two groups of "true" worms are common or abundant in the Appalachians and Ozarks. The annelids (phylum Annelida) are the more familiar group, while the nematodes (phylum Nematoda) are the more numerous with respect to the number of species and the number are individuals.

The best-known annelid is the earthworm. Earthworms are soil-inhabiting animals and thus out of sight much of the time. However, they come to surface at night to feed (the basis for one of their common names), where they are exposed to numerous predators, including salamanders, frogs, various birds, snakes, small mammals, and even some other invertebrates such as beetles, snails, and slugs. Earthworms also can be found on the surface of the ground during the day after a heavy rain. As is generally known, earthworms play an important role in nature by breaking up large pieces of decaying organic matter (e.g., dead leaves) and, as a result of their feeding activities, convert this into humus, the dark brown or black material that makes up the organic component of soil. Moreover, the passageways that earthworms make help to aerate the soil and increase the infiltration of water.

The most familiar species of earthworm is the nightcrawler (*Lumbricus terrestris*), a reddish-brown worm that can exceed 20 cm in length. It is commonly used as a bait for fish and is almost invariably studied in introductory biology classes. Although a common and familiar animal, the nightcrawler is not indigenous to North America. It was introduced from Europe, probably by English settlers in the early seventeenth century, and has since spread to most of the continent.

The inadvertent introduction (probably most often in potted plants) of species of earthworms to the United States from Asia and Europe has become a major problem because their presence in the soil at a particular site has a negative effect on soil structure and nutrient cycles (Bohlen et al. 2004). These earthworms break up decaying organic matter and spread it though the soil. This increases the leaching of soil nutrients and reduces the accumulation of litter on the soil surface. This affects the survival of seedlings of trees and other plants, potentially leading to a loss of biodiversity and changes in community composition. Invasive earthworms have been an especially acute problem for forests at higher latitudes in North America, where native species of earthworms were absent.

To say that nematodes are abundant in nature would be an understatement. Nematodes are ubiquitous in all types of aquatic and terrestrial habitats, where they usually outnumber all other animals present in both number of individuals and number of species. One teaspoon of forest soil and humus typically contains several hundred nematodes, and the number present in an entire forest is simply inconceivable. Many nematodes are free-living, but others are parasites or pathogens of plants and animals, and some can be economically important. For example, root-feeding nematodes may sometimes kill young plants. As a group, these worms, the great majority of which are less than 1 mm in length and many of which are microscopic, play an important role in the decomposition of organic matter and the recycling of nutrients in soil. Some species feed on bacteria (and thus are bacterivores), others on fungi (and thus are fungivores), while still others are predators of other nematodes and other tiny invertebrates. In turn, many other organisms, including some nematode-trapping fungi, make use of nematodes as a food source. Nematodes are an important part of the biota of forests and other ecosystems. These organisms are active only in the presence of at least some moisture, and they commonly transform into highly resistant structures called cysts when their surroundings dry out.

8.4.5 Tardigrades

The tardigrades (phylum Tardigrada) are the last group considered in this book. They are included not because of their ecological importance but because they are among the most fascinating animals found in the forests of the Appalachians and

Ozarks. Tardigrades are among the most resilient animals known to science and are able to survive under extreme environmental conditions that would kill almost any other type of organism. Although tardigrades can survive under severe conditions for centuries, they require the presence of at least a film of moisture in order to be biologically active. In a forest, they are most commonly associated with bryophytes (especially mats of mosses), lichens, and leaf litter. An individual tardigrade is very small (body length of only 0.1–1.5 mm), has a bilaterally symmetrical body, and four pairs of legs ending in claws. These organisms mostly feed on fluids from plants and the cells of algae, but some species are carnivorous and consume other small creatures, including other tardigrades. Tardigrades are commonly called water bears. The name reflects the way they walk, which early observers thought was similar to that of a bear.

Many biologists and non-biologists have found tardigrades to be interesting animals, but serious studies of their biodiversity are lacking in most regions of the world, including the Appalachians and Ozarks. There have been several papers published on the tardigrades of the Ozarks. Examples include Hidalgo and Coombs (1985), who reported nine species from Missouri, and Land et al. (2012), who recorded eight species from Arkansas. However, data from the ATBI provide some indication of what an intensive survey could yield. Eighty-five species of tardigrades have been recorded from the Great Smoky Mountains National Park, and 26 of these were new to science. Tardigrades simply represent another element of the biota of forest ecosystems that is still rather incompletely known.

References

Bohlen PJ, Scheu S, Hale CM, McLean MA, Migge S, Graffman PM, Parkinson D (2004) Non-native invasive earthworms as agents of change in northern temperate forests. Front in Ecol Environ 2:427–435

Discover Life in America (2024) How many kinds of life inhabit Great Smoky Mountains National Park? https://dlia.org/smokies-species-tally/

Dourson DC (2015) Land snails of West Virginia. Goatslug Publications, Bakersville, NC

Gaddy LL (1986) Twelve new ant-dispersed species from the southern Appalachians. Bull Torrey Bot Club 113:247–251

Hidalgo H, Coombs D (1985) Tardigrada from Missouri. Trans Kans Acad Sci 88:121–134

Hubricht L (1972) The land snails of Arkansas. Sterkiana 46:39–45

Hyseni C, Garrick RC (2019) Ecological drivers of species distributions and niche overlap for three subterranean termite species in the Southern Appalachian Mountains, USA. Insects 10:33. https://doi.org/10.3390/insects10010033

Johnson N, Triplehorn C (2004) Introduction to the study of insects, 7th edn. Cengage Learning, Boston, MA

Karnish A (2024) Primers in the plant sciences. Seed dispersal by ants: a primer. Int J Plant Sci 185(5):403–411. https://doi.org/10.1086/730787

Land M, Musto A, Miller WR, Starkey DE, Miller JD (2012) Tardigrades of the University of Central Arkansas Campus, Conway, AR. Southeast Nat 11:469–476

Lessard J-P, Dunn RR, Sanders NJ (2007) Rarity and diversity in forest ant assemblages of the Great Smoky Mountains National Park. Southeast Nat 1:215–228

Oesch RD, Watrous LE, Barnhart MC (2013) Land Snails and Slugs of Missouri. Missouri Department of Conservation, Jefferson

Sharkey MJ (2001) The all taxa biodiversity inventory of the great smoky mountains national park. Fla Entomol 84:556–564

Spencer LA (2006) Arkansas butterflies and moths. Ozark Society Foundation, Little Rock, AR

Wheeler WC, Whiting M, Wheeler QD, Carpenter JM (2001) The phylogeny of the extant hexapod orders. Cladistics 17:113–169

Chapter 9
Fungi of the Appalachians and Ozarks

Abstract The fungi are among the least known and least appreciated components of forest ecosystems. The vast majority of the species present are microfungi and cannot be easily observed directly in nature, thus usually escaping our attention. However, fungi are absolutely essential for the functioning of the forest ecosystem. Saprotrophic fungi play an important role in regulating natural processes, including the decomposition of dead plant material. All trees are dependent upon the mycorrhizal associations they establish with fungi, while parasitic and pathogenic fungi reduce the fitness of their plant hosts, including trees, sometimes resulting in the death of the host. Macrofungi produce fruiting bodies that are visible to the naked eye, and these fruiting bodies are fed upon by a wide variety of animals, and those of some species are collected and eaten by humans. Although most fruiting bodies are ephemeral and do not persist more than a few days, many are conspicuous enough to be easily noticed. The ecological roles and different types of fungi associated with the forests of the Appalachians and Ozarks are discussed in this chapter.

Keywords Ascomycetes · Basidiomycetes · Classification · Groups of fungi · Microfungi · Mycorrhizal fungi · Parasites · Saprotrophs · Classification

9.1 Fungi as Organisms

It would be difficult to find someone who wouldn't recognize a mushroom, at least the kind commonly found in supermarkets and on salad bars. Most people would have heard that some mushrooms are poisonous and thus should be avoided. However, few people truly understand the importance of mushrooms and other fungi in forests (Cripps 2004). It would not be an exaggeration to say that there wouldn't be forests without fungi. The purpose of this chapter is to describe the

S. L. Stephenson, R.-M. Muzika, *Forests of the Southern Appalachian Highlands and Interior Highlands of the United States*, Plant and Vegetation 20, https://doi.org/10.1007/978-3-032-01155-8_9

biological and ecological roles that fungi have in the forests of the Appalachians and Ozarks and the types of fungi involved.

The fungi (singular: fungus) are members of the kingdom Fungi. The number of different species of fungi is still unknown, but the best current estimate places the number well into the millions. However, only about 150,000 of these have been formally described (Hyde 2022). Fungi are found in every habitat on the earth, although they are poorly represented in marine habitats. The vast majority of fungi are microscopic and escape notice in the field except under special circumstances. This is not the case for those fungi that produce macroscopic fruiting bodies, some of which are typically referred to as mushrooms.

9.1.1 Saprotrophs and Parasites

In some ways fungi are similar to plants and were once considered members of the plant kingdom, but the two groups differ in several major respects. Unlike plants, fungi lack chlorophyll and cannot produce their own food through photosynthesis. Instead, they obtain their food by breaking down dead organic matter, or in some cases, by attacking and living on or within living plants, animals or even other fungi. Fungi that depend upon dead organic matter as their food source are called saprotrophs, while those that feed on living hosts are referred to as parasites if the host is harmed but not killed and pathogens if their presence produces a condition (called a disease) that has the potential of causing the death of the host. The distinction between parasite and pathogen is not absolute, and a parasite may become a pathogen over time or under a different set of circumstances (Stephenson 2010).

As a result of their activities, saprotrophic fungi play a very important role in regulating natural processes. For example, it has been estimated that in a year several million leaves fall to the ground in each 0.4 ha of a temperate deciduous forest of the type common throughout the Appalachians and Ozarks. These leaves do not accumulate year after year in part because of the saprobic activity of various fungi. As a result, essential nutrients in the leaves are recycled to the soil. Fungi are also the major group of organisms responsible for wood decay. Although the vast majority of fungi are terrestrial, there are a number of aquatic forms, and some of these are of considerable ecological significance. For example, in small, well-aerated streams in the Appalachians and Ozarks, aquatic fungi play a key role, not only in the decomposition of organic matter introduced into streams but also as intermediates in food chains involving many aquatic insects and other invertebrates.

9.1.2 Mycorrhizal Fungi

Parasitic and pathogenic fungi obviously reduce the fitness of their plant host to the point that the latter can die, but there is another plant-fungus relationship in which the plant benefits. This relationship, known as a mycorrhizal association, is mutually beneficial to both the plant and the fungus (i.e., it is symbiotic). Mycorrhizal associations, which are restricted to certain groups of fungi, involve the formation of intimate associations between these organisms and vascular plants, including all forest trees. Mycorrhizal associations or mycorrhizae (literally "fungus roots") are exceedingly widespread and known to occur in perhaps 80–90% of all vascular plants. In brief, the fungus benefits the plant by increasing its ability to take up water and nutrients such as nitrogen, phosphorus, potassium, and calcium from the soil, while the plant benefits the fungus by supplying it with organic molecules produced as a result of photosynthesis. Traditionally, mycorrhizal associations formed between vascular plants and fungi have been divided into two types on the basis of structural differences in the way the association is formed. In an ectomycorrhizal (ECM) association, most of the fungal hyphae (described below) are on the outside of the root, where they form a distinct layer called a mantle (Fig. 9.1). Some hyphae extend into the root, but they do not disrupt the root cells. In the second type, referred to as arbuscular mycorrhizae (AM), most of the hyphae are inside the root, where they actually penetrate root cells (Fig. 9.2). Their name refers to the structures (arbuscules) that the hyphae of these fungi form inside the cells of the root.

Of the two types, AM fungi are by far the more common. Most herbaceous plants are AM and so are many of the trees that make up the forests of temperate and tropical regions of the world. These include members of the families Cupressaceae and Taxaceae among the gymnosperms and the angiosperm families Rosaceae, Fabaceae, Oleaceae, and Tiliaceae (Policelli et al. 2020). Common AM trees in the forests of the Appalachians and Ozarks include black locust (*Robinia pseudoacacia*), magnolia (*Magnolia*), cherry (*Prunus*), maple (*Acer*), ash (*Fraxinus*), and sourwood (*Oxydendrum arboreum*). A few trees can form both ECM and AM associations. Two examples found in both the Appalachians and Ozarks are birch (*Betula*) and alder (*Alnus*).

Only about 3% of all vascular plants form ECM associations, but some of the plants that do (mostly trees) are widespread. A single family of gymnosperms (the Pinaceae) is ECM, but it includes several common and widely distributed genera

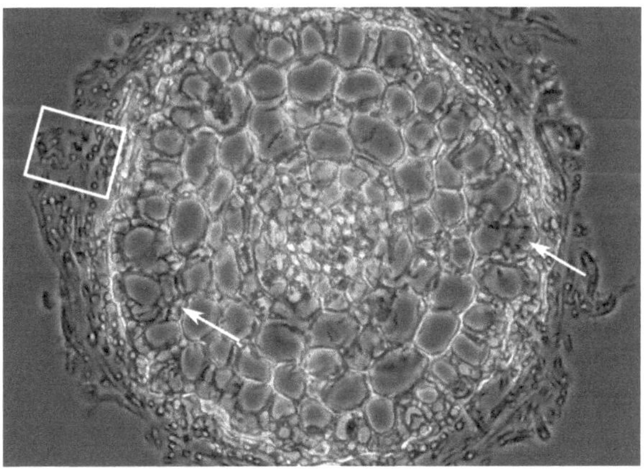

Fig. 9.1 Section of a root from a tree that forms an ectomycorrhizal association with a fungus. The position of the mantle is indicated by the small square, while the arrows show the fungal hyphae found within the tissues of the root. (Photo by Steven L. Stephenson)

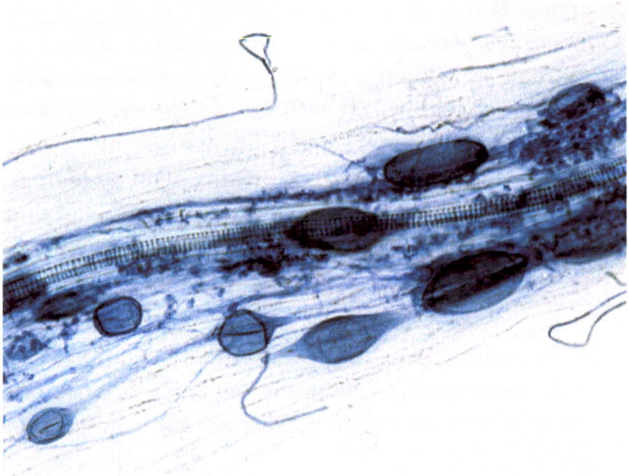

Fig. 9.2 A portion of a root with a AM mycorrhizal fungus present. The large dark structures are vesicles. (Photo by Elizabeth Koziol)

including *Abies* (fir), *Larix* (larch), *Picea* (spruce), *Pinus* (pine), and *Pseudotsuga* (Douglas fir). The angiosperms contain a number of ECM-forming genera, including *Betula* (birch), *Quercus* (oak), *Fagus* (beech), and *Castanea* (chestnut and chinquapin). In many places in both the Ozarks and Appalachians ECM trees dominate the forests. However, the ECM-forming taxa of fungi are not the same for all species of ectomycorrhizal trees. Some species of fungi can form ECM associations with a wide range of host trees but others may form these associations with only a few host trees. Clearly there is much still to be learned about mycorrhizal relationships.

Although the assemblages of ECM fungi found in the forests of the Appalachians and the Ozarks are likely to have considerable overlap (Elliott and Stephenson 2018), the extent to which this is the case remains unknown because the fungi of the Ozarks are not especially well-documented, since only limited surveys have ever been carried out (Swartz 1933; Ben Hassine Ben Ali et al. 2018). The same is true for all other groups of fungi (Alanbagi et al. 2019).

9.1.3 Studies of Fungi

The mycologists Ron Peterson and L. R. Hesler at the University of Tennessee, John Couch and William Coker at the University of North Carolina, and Orson Miller at Virginia Tech carried out studies of the macrofungi of the Appalachians during their careers (e.g., Coker 1923; Hesler 1960; Petersen 1971; Coker and Couch 1974; Miller 1977). All three universities are either within or occur in close proximity to the Appalachians, and a number of graduate students of these five professors earned graduate degrees from research carried out in the region. The research these students carried out yielded a considerable number of publications on many different groups of fungi. In 1998, the Great Smoky Mountains National Park Smokies began a park-wide biological inventory, referred to as the All Taxa Biodiversity Inventory

(ATBI), of all life forms, including fungi. This effort, which is still ongoing, has yielded a tremendous body of data on fungi for this portion of the southern Blue Ridge. In general, the fungi are less known for the Ozarks than for the Appalachians, although a number of recent studies (Vasilyeva and Stephenson 2009; Ben Hassine Ben Ali et al. 2018; Alshammari and Stephenson 2018; Alanbagi et al. 2019; Alshammari et al. 2020) have expanded the information available for several different groups of fungi. Other recent studies (Bruhn et al. 2000; Lee et al. 2017) have been directed towards pathogenic fungi. A number of species of macrofungi have been described from specimens collected in the region. Perhaps the most spectacular example is a red species of *Amanita* (*A. arkansana*) described from Arkansas by Rosen (1926).

9.1.4 Structure and Classification of Fungi

The basic structure of fungi is quite different from that of plants. The vegetative body of all but the simplest fungi is made up of a system of very finely branched, microscopic (only 0.5–3.5 μ in diameter), threadlike structures called hyphae (singular: hypha). The entire system of hyphae (the mycelium) typically occurs in soil, leaf litter, or decaying wood, where the individual hyphae obtain the nutrients and water the fungus needs to grow. As they grow, the hyphae spread over or penetrate the substrate with which they are associated. More often than not, the hyphae occur within the substrate and thus are not readily apparent in nature. The mycelium can be so extensive and the collective surface of the hyphae so large that it can invade and affect a considerable volume of biomass or soil. For a fungus involved in a mycorrhizal association, this increases the amount of nutrients that it takes up from the soil and then transfers to the host plant. Saprotrophic fungi synthesize exoenzymes inside the hyphae that are then secreted outside the hyphal wall, where their function is to break down complex macromolecules into smaller units to be taken up by the hyphae for growth. This allows these fungi to have the potential to quickly decompose organic substrates such as leaf litter or wood.

After a period of growth and under favorable conditions of temperature and moisture, a mycelium usually produces one or more fruiting structures. For the majority of fungi (the microfungi), these structures are not visible with the naked eye unless they occur in fairly dense masses on an exposed substrate. For certain other fungi (the macrofungi), these larger structures (or fruiting bodies) are typically visible and sometimes reach a considerable size. Fruiting bodies occur in a wide variety of different colors, shapes, and sizes. The familiar mushroom is recognized as having an initially inverted bowl-shaped cap held aloft on a stem-like stalk (also called a stipe). It should be noted that the term mushroom is often restricted to those fungi that have blade-like structures (gills) beneath the cap that radiate outwards from the top of the stalk. However, the term has sometimes been extended to encompass all macrofungi. To avoid possible confusion, the macrofungi with gills are better referred to as agarics or gilled mushrooms (Alexopoulos et al. 2007). The

Fig. 9.3 The oak-loving gymnopus (*Gymnopus dryophilus*) is a very common agaric associated with decomposing oak leaves and twigs in the forests of the Appalachians and Ozarks. (Photo by Steven L. Stephenson)

oak-loving gymnopus (*Gymnopus dryophilus*) is a very common agaric that decomposes oak leaves and twigs in the forests of the Appalachians and Ozarks (Fig. 9.3).

With the advent of molecular techniques, the classification system used for fungi has undergone major changes that have resulted in some species thought to be closely related assigned to entirely different groups (Tedersoo et al. 2018). However, these same techniques have made it possible to identify fungi to a far greater extent than previously possible. In the past, macrofungi were identified largely on the basis of the morphology of the fruiting body and, for agarics and some non-agarics, the colors of the spores produced. Since it is not always possible to use molecular techniques to identify a particular specimen, especially in the field, traditional methods of identification based upon morphology are still useful.

The vast majority of the fungi that form ectomycorrhizal associations belong the division Basidiomycota (commonly known as basidiomycetes). Only a few belong to a second division, the Ascomycetes (commonly known as ascomycetes). The basidiomycetes include many of the better-known genera of macrofungi, including the agarics *Amanita* and *Russula*. In contrast, the fungi that form arbuscular mycorrhizae belong to a third division, the Glomeromycota (commonly known as glomeromycetes). These are microfungi not visible to the naked eye.

9.2 Groups of Fungi and Their Roles

9.2.1 Basidiomycetes

Most basidiomycetes can be assigned to one of number of different morphological groups. We now know that not all of the fungi placed in a particular morphological group are always closely related taxonomically, but being able to recognize these groups still represents an effective way of developing a better understanding of the biological and ecological roles the members of a particular group play in forest ecosystems.

The gilled mushrooms or agarics, the group mentioned above, display a considerable diversity with respect to color, size, shape, and their role in nature. Many species form mycorrhizal associations with forest trees, while others decompose wood or forest floor leaf litter. Invariably, one of the first questions asked about any large fungus, especially an agaric, is whether it is edible. Although many fungi are, others are deadly poisonous. Unfortunately, there is no simple rule or test to distinguish edible from poisonous fungi. The best approach is to be exceedingly careful, which means not collecting any fungus for the table unless it has been identified with absolute certainty.

Chanterelles are funnel- or trumpet-shaped fungi with caps that are slightly to deeply depressed in the center. Chanterelles resemble agarics, but the cap is not clearly distinct from the stalk, and the spore-bearing surface is quite different. Some chanterelles have a cap that is nearly smooth underneath, while in others the lower surface has a network of wrinkles or ridges that also extend for some distance down the stalk. The ridges have many forks and cross-veins and are always blunt-edged in contrast to the sharp-edged, blade-like "true" gills of an agaric. True chanterelles form mycorrhizal associations with forest trees. The fruiting bodies of chanterelles can often be found in oak-dominated forests during the summer, where "mushroom hunters" collect these edible fungi for human consumption.

The boletes are very similar in overall appearance to agarics. The fruiting body has a cap and a stalk, but instead of having gills, the lower surface of the cap is composed of a spongy layer containing numerous tubes, on the inner walls of which the spores are produced. The openings to the tubes are referred to as pores. Boletes are an ecologically important group of fungi because they form mycorrhizal associations with forest trees. Some species are edible but others are not.

The polypores are like boletes in that they have pores beneath the cap. However, while the fruiting body of a bolete is soft and fleshy, that of a polypore is usually tough, corky, or woody. The turkey tail (*Trametes versicolor*) is an example of a polypore that is a very common wood-decomposing fungus in the forests of the Appalachians and Ozarks (Fig. 9.4). In contrast to boletes, which grow on the ground, most polypores grow on decaying wood. This reflects a fundamental difference in their biology. Most polypores are decomposers of woody debris on the forest floor, but an appreciable number occur as root- and heart-rots on living trees, and a few are mycorrhizal. In contrast, boletes form mycorrhizal relationships with forest trees. Because the fruiting bodies of most polypores are tough or hard, they can persist in nature for months or even years. Their fruiting bodies are conspicuously present in the forests of the Appalachians and Ozarks in every season of the year. Since their fruiting bodies often project outward from the tree, log, or stump on which they happen to occur, polypores are often referred to as shelf fungi or bracket fungi.

Crust fungi are the least known of the more abundant groups of fungi found in all types of forests. These fungi produce fruiting bodies that are more or less flat and closely adherent to the surface of the decaying log or other woody debris on which they grow. In many instances this is the bottom surface, and the presence of the fungus is revealed only if the log or piece of woody debris is rolled over. The

Fig. 9.4 The turkey tail fungus (*Tremetes versicolor*) is one of the most common polypores in the forests of the Appalachians and Ozarks. The zonation of the cap makes this fungus easy to recognize. (Photo Steven L. Stephenson)

fruiting bodies of some crust fungi are so thin and closely adherent that they have the same general appearance as a coat of colored paint. The upper exposed surface of the fruiting body is the spore-bearing surface, and it may be smooth, wrinkled, or composed of small pores or spines. Most crust fungi are hard or tough, but a few are somewhat gelatinous. Like a number of other groups of basidiomycetes, the crust fungi are now known to be rather heterogeneous, and the traditional taxonomy used for these fungi lumped together species that were not at all closely related. The crust fungi are mostly decomposers of wood, but some species are mycorrhizal and a few are pathogens.

The tooth fungi (sometimes also called spine fungi) are another group of basidiomycetes that share a morphological feature (a fruiting body with tooth- or spine-like projections on the lower surface) but are now known to represent a heterogeneous taxonomic assemblage. Some tooth fungi form mycorrhizal associations, while others decompose wood. Both groups contain examples that are edible.

Coral fungi are characterized by fruiting bodies typically consisting of one or more upright branches, sometimes intricately divided. The spore-bearing surface is located on these branches. Coral fungi resemble the corals found in the ocean, hence their name. Fungi that have been assigned to the group are not all closely related, something that has become apparent only recently as a result of molecular studies. Some species form mycorrhizal associations with forest trees, while others are decomposers of wood and litter.

In contrast to all the groups of basidiomycetes mentioned thus far, puffballs produce their spores inside the fruiting body and not on some portion exposed directly to the atmosphere. Spores are released through an opening that develops at the top of the fruiting body or as a result of the breaking apart of the upper portion of the fruiting body (or sometimes the entire fruiting body). Most puffballs are no larger than a golf ball, but a few species have fruiting bodies that can reach the size of a soccer ball or basketball. True puffballs (since there other fungi that can resemble puffballs) are edible when young. Puffballs are decomposers of litter, organic matter in the soil, or (more rarely) wood. They occur during late summer and fall in most forests of the Appalachians and Ozarks. The most species most likely to be encountered is the gem-studded puffball (*Lycoperdon perlatum*), a decomposer of litter and humus (Fig. 9.5).

Fig. 9.5 The gem-studded puffball (*Lycoperdon perlatum*) occurs on the forest floor throughout both the Appalachians and Ozarks. While it is still immature, this fungus is edible. (Photo by Steven L. Stephenson)

Earthballs are often confused with puffballs, since they superficially resemble puffballs and grow in similar habitats, but earthballs have a much thicker and tougher rind-like outer wall, and the spore mass on the inside is purplish-black when the fungus is mature, which is darker than the spore mass in puffballs. More importantly, earthballs form mycorrhizal associations with forest trees, while (as noted above) puffballs are decomposers of organic matter in the soil. Although members of the two groups are similar in appearance, they are not closely related.

9.2.2 *Ascomycetes*

The second major taxonomic assemblage of fungi, the ascomycetes, includes some familiar macrofungi, including the truffles, morels, and the larger cup fungi. Most ascomycetes are microfungi, and some can be deadly plant pathogens. Two examples are the fungus (*Bretziella fagacearum*) that causes oak wilt, which occurs in both the Appalachians and Ozarks, and the fungus (*Cryphonectria parasitica*) that resulted in the demise of the American chestnut in the Appalachians. There are far more species of ascomycetes than basidiomycetes, but few of these produce fruiting bodies that can be readily observed under field conditions. Various types of cup fungi are the ascomycetes most likely to be noticed in nature, although many species are small enough that they are easily overlooked. Cup fungi have a fruiting body that is shaped like a cup, shallow bowl, or urn, with the spores produced on the inner surface layer of this structure. The fruiting bodies of many species of cup fungi are brightly colored and often grow in groups or clusters on decaying wood or the surface of moist, humus-rich soil. The brightly colored fruiting bodies of the scarlet cup (*Sarcocoscypha occidentalis*) occur on hardwood twigs in the forests of the Appalachians and Ozarks in mid-summer to early autumn (Fig. 9.6). Most cup fungi are decomposers of wood or other types of plant debris.

The flask fungi derive their name from the shape of their microscopic fruiting bodies. Numerous and densely packed fruiting bodies typically grow within a hard,

Fig. 9.6 The scarlet cup (*Sarcoscypha occidentalis*) is a small but rather distinctive cup fungus commonly occurring on twigs in forests of the Appalachians an Ozarks. (Photo by Steven L. Stephenson)

sterile structure called a stroma (plural: stromata), and it is the entire stroma that is readily apparent in nature. The stromata of flask fungi exhibit a wide range of different forms. Some are clublike structures, a few are more or less hemispherical, and numerous examples occur as a cushion-like layer over the surface of a decaying log or stump. Many common flask fungi are decomposers of wood, but others are parasites or virulent pathogens of particular plants. For example, the chestnut blight fungus that virtually eliminated the American chestnut from the forests of eastern North America is a flask fungus.

The fruiting bodies of dead man's fingers are clublike structures, cylindrical to somewhat irregular in shape, 2 to 10 cm tall and 1 to 2 cm thick, and powdery white to dirty white when young but becoming black when mature. The typically finger-like appearance of this fungus is the basis of the common name. Fruiting bodies are found on decaying logs and stumps of deciduous trees, where they often grow in small clusters. When a mature fruiting body is broken apart the pure white interior is revealed.

The best known ascomycetes easily recognized in the field are the morels (also called sponge mushrooms), the most commonly harvested wild edible macrofungi in North America. The fruiting season for morels is in the spring, generally during April and May. During this period "morel hunters" search for their fruiting bodies in forested areas. Morels have a fruiting body that consists of a stalk and an upper portion (often referred to as a "head") with distinctive ridges and pits. Morels are generally regarded as decomposers of litter and humus, but the consistent association of certain morels with particular species of trees suggests that they sometimes form mycorrhizal associations with forest trees.

The yellow morel is the most common morel in eastern North America, although the category is now known to encompass a complex of several species that are morphologically very similar. Yellow morels have a total height of 5 to 25 cm. The cap is yellow-brown to tan, with the pits in the cap somewhat lighter in color. The stalk is hollow and white to pale yellow-brown. These fungi are found on the ground in deciduous forests, where they are solitary or grow in small to large groups. In truly exceptional instances it is possible to find more than several hundred fruiting bodies in an area of less than a hectare. The number of fruiting bodies varies considerably from year to year, but in some years they can be abundant. Although edible, morels should always be cooked before they are consumed.

9.3 Edible Fungi

As noted above, it is very well known that the fruiting bodies of some species of macrofungi are edible and collected for food by humans. However, the fruiting bodies of fungi represent an even more important food resource for forest animals, including small mammals as well as some larger forms such as the white-tailed deer. The fruiting bodies that these animals feed upon are not always evident. Hypogenous (underground) fruiting bodies have evolved in several groups of fungi, including both ascomycetes and basidiomycetes. The best-known examples are truffles, which can occur in association with such trees as beech, oak, and pine. Much more common hypogenous fungi in forests are the false truffles, with members of the genus *Elaphomyces* especially common in the forests of eastern North America. The fruiting bodies of *Elaphomyces* range from 1 to 3 cm in diameter and often occur in clusters (Fig. 9.7). The white-tailed deer is known to dig up and consume these fruiting bodies (which are sometimes referred to as "deer truffles"), and they are an important item in the diet of the northern flying squirrel (Stephenson 2010).

No discussion of forests would be complete without mentioning edible macrofungi. As indicated previously for a couple of specific examples, forests represent fertile hunting areas for fungi collected for human consumption. In the spring, morels (especially black morels such as *Morchella angusticeps* [Fig. 9.8]) and yellow morels such as the tulip morel (*M. diminutiva*) or the common morel (*M. americana*) are the primary targets, but during the summer and fall, attention is directed towards various other fungi, including such notable examples as the oyster mushroom (*Pleurotus ostreatus*), the sulphur-shelf polypore (*Laetiporus sulphureus*) (Fig. 9.9), honey mushroom (*Armillaria mellea*), hen of the woods (*Grifola frondosa*), shaggy mane (*Coprinus comatus*), and lion's mane (*Hericium erinaceus*) along with several species of boletes (*Boletus* spp.) and chanterelles (*Chantarellus* spp.). Almost all of these are wood-decay or ectomycorrhizal fungi (Elliott and Stephenson 2018).

Fig. 9.7 Hypogeous fungi such as the deer truffle (*Elaphomyces americanus*) are an important food resource for forest mammals, including the white-tailed deer. (Photo by Todd Elliott)

Fig. 9.8 The black morel (*Morchella angusticeps*) is widespread throughout the Appalachian and Ozarks. It can be distinguished from the common morel (Fig. 13.9) by its much darker color. (Photo by Steven L. Stephenson)

Fig. 9.9 The sulphur-shelf polypore (*Laetiporus sulphureus*) is one of the prized edible fungi found in the Appalachians and Ozarks. (Photo by Steven L. Stephenson)

9.4 Ecological Roles of Fungi

9.4.1 Mycorrhizal Associations

One would expect that those fungi that form mycorrhizal associations with the roots of trees would occur on the forest floor. However, many of the fruiting bodies found on the forest floor are not those of ectomycorrhizal fungi. Saprotrophic fungi that decompose leaf litter, woody twigs, and various other types of plant debris also produce macroscopic fruiting bodies that occur in the same habitat. The fruiting bodies of ectomycorrhiza fungi arise from a mycelium in the ground, while those of saprotrophic fungi arise from mycelia that occur within or on the particular substrate they are decomposing. However, in many instances this distinction is not immediately obvious if one is walking through a forest.

Common genera of saprotrophic associated with decaying leaves are *Gymnopus* and *Marasmius*, while *Microstroma* and *Mycena* decompose small twigs. A few of the saprophytic fungi have an amazing specificity for the substrates they decompose. For example, the sweet gum fruit fungus (*Xylaria liquidambar*) is found only on old decaying fruits of *Liquidamber*, and the magnolia-cone xylaria (*Xlyaria magnoliae*) is found only on old decaying fruits of species of magnolia.

9.4.2 Decomposers

All trees (with some unusual exceptions such as palms) have secondary growth and thus produce wood (i.e., secondary xylem). The fungi that decompose wood (also referred to as coarse woody debris) have the same fundamental role, but the process of decomposition takes place over various periods of time. The main trunk of a fallen tree may require several decades to decompose completely, and some fallen American chestnut logs were still present well over half a century after they first fell to the forest floor after being killed by the chestnut blight fungus. By definition, the term wood refers to the secondary xylem of vascular plants that have secondary growth. All trees (with a very few exceptions such as palms) have secondary growth and thus produce wood. Herbaceous plants generally do not undergo secondary growth. Although some herbaceous plants (e.g., *Smilax*) can have tough stems, the stems are not actually woody. Fungi that decompose coarse woody debris can be classified into two groups based on the component of wood they specialize at breaking down. "Brown-rot" fungi digest cellulose but leave lignin, which has a brown color, intact. The destruction of cellulose results in cubical fracturing of the wood, creating richly colored brown cubes of wood that mix into soil and humus. The sulphur-shelf mushroom is a brown-rot fungus that can invade dead broadleaf trees or weakly parasitize mature but still living ones. The sulphur-shelf mushroom is very well known because it is an easily recognized edible fungus. Other brown-rot fungi are Dyer's polypore (*Phaeolus schweinitzii*), red-belted conk (*Fomitopsis mounceae*), and the birch polypore (*Piptoporus betulinus*). Most brown-rot fungi are associated with conifers.

"White-rot fungi" specialize at digesting lignin, leaving the remaining wood pale and fibrous. Common examples are the tinder fungus (*Fomes fomentarius*), the artist's conk polypore (*Ganoderma applanatum*), the turkey tail mushroom (*Trametes versicolor*), the red ring rot fungus (*Porodaedalea pini*), and the oyster mushroom. The latter is a widely sought edible. Interestingly, the fungi that decompose the bark of a tree are not necessary the same taxa that decompose wood. One reason for this relates to the fact that bark is a higher quality substrate (i.e., higher levels of most nutrients) for decomposers than wood.

In a living tree, a major portion of the trunk consists of dead xylem cells and thus are subject to being exploited by wood-decay fungi. Sometimes the entire center of the trunk can be decomposed, resulting in a still living but hollow tree. The presence of wood-decay fungi in a living tree is often difficult to detect unless they produce fruiting bodies on the outside of the trunk (Stephenson 2010).

Although often not considered when the fungi associated with the forests of temperate North America are discussed, there are particular fungi that decompose the dung of herbivores. Dung actually consists of undigested plant material to which animal products have been added. The exact nature of dung depends upon the type of animal from which it was derived and the original type of plant material consumed by the animal. Dung tends to decompose very rapidly because it is already broken up to some extent and also contains high levels of nitrogen and is typically moist. The fungi associated with dung (referred to as coprophilous fungi) are very diverse and include taxa not found on any other substrates in forests. A few macrofungi fruit on dung, including some members of the genus *Coprinus* (considered in the traditional sense).

9.4.3 Parasitic and/or Pathogenic Fungi

Fungi that are parasitic and/or pathogenic (since some species that are initially parasitic can become pathogenic) fungi represent a major component of any forest (or any other) ecosystem. Various rusts and smuts (two groups of pathogenic basidiomycetes) infect numerous types of herbaceous plants but largely go unnoticed because the fruiting structures they produce are small. One notable exception is mayapple rust (*Allodus podophylli*), which manifests itself as conspicuous orange splotches on the leaves.

Rust diseases are very common and relatively easy to observe in nature because of the colorful spores they produce. As a group, rusts are among the most problematic of all fungi because of their complex life cycles that often require multiple hosts to complete and also because they are obligate parasites (i.e., restricted to specific hosts). Rust diseases are widespread in forests as well as at the urban/forest interface. Notable examples include pine oak gall rust (*Cronartium quercuum*) and cedar apple gall rust (*Gymnosporangium juniperi-virginianae*). White pine blister rust

(*Cronartium ribicola*) can cause mortality in all five-needle pines and is a serious problem in the western United States and in a few areas in the eastern United States. It may become a greater issue for white pines in the east. Other fungal diseases of trees important in the western coniferous forests have the potential of becoming increasing important in the Ozarks. One example is the root disease fungus *Heterobasidion* (Lee et al. 2017).

Threats to forests will be addressed in a subsequent chapter, but it is worth noting that a vast number of forest health issues in the Appalachians and Ozarks result from fungi. For example, *Armillaria mellea* (honey mushroom) has long been cited as an important secondary mortality agent in hardwood forests of the Appalachians (Wargo 1977). Species of *Armillaria* are widespread and more recent studies in the Missouri Ozarks have provided insight into the ecological roles of these fungi (Bruhn et al. 2000; Lee et al. 2021) and their associations with other mortality agents in the Ozarks (Kelley et al. 2009).

9.5 Ecosystem Services

Directly or indirectly, fungi provide a number of what are referred to as ecosystem services to forests. The presence of wood-decay fungi in the trunks of still standing dead trees predisposes the wood for cavity-excavating birds such as woodpeckers (Elliott et al. 2019). As such, the birds are largely dependent upon the fungi (mostly polypores), although this relationship is often overlooked. The fruiting bodies of larger polypores provide shelter and a breeding place for numerous insects (Jonsell and Nordlander 2004). For example, the often quite large (sometimes as much as half a meter across) old and decaying fruiting bodies of the artist's conk polypore usually support an abundant and taxonomically diverse assemblage of insects as well as non-arthropod invertebrates such a mites and nematodes (Wheeler and Blackwell 1984).

By weakening the structural integrity of the trunk and roots, wood-decay fungi contribute to the windthrow of trees in forests. Once on the forest floor, the decomposing trunk provides a habitat for numerous other organisms, including bryophytes, insects, small mammals, and other fungi. When large pieces of wood are present in a small stream, the entire stream ecology is affected, which affects an entirely different assemblage of organisms.

The oomycetes (commonly referred to as water molds) are not true fungi, although they resemble fungi in a number of respects. However, some oomycetes are important pathogens of trees and other plants. Damage to forests suspected to have been caused by the genus *Phytophthora* was first detected in the United States two centuries ago, and this organism probably contributed the death of American chestnut trees prior to the arrival of the chestnut blight fungus. It still remains an important factor in oak mortality in both the Appalachians and Ozarks.

Fig. 9.10 Fruiting bodies of a slime mold on the bark of a decaying log. These organisms are common in both the Appalachians and Ozarks. (Photo by Steven L. Stephenson)

9.6 Myxomycetes

Since one of the coauthors of this book has studied myxomycetes (slime molds) over much of his career, it would be remiss not to mention these organisms, for whom forests are their primary habitat. Myxomycetes were once thought to be fungi but actually belong to a totally different phylum (the Amoebozoa). The vegetative (or amoeboflagellate) stage in the myxomycete life cycle is a fundamental element of the soil biota, where these organisms play an essential but very much underappreciated role in nutrient cycling. Myxomycetes consume bacteria and thus make available nutrients locked up in bacterial biomass that would otherwise be unavailable to plants. Reproduction in myxomycetes involves the formation of fruiting bodies, many of which, although small, are visible to the naked eye. These fruiting bodies are often colorful and sometimes quite ornate (Fig. 9.10). A close examination of decaying wood, especially stumps and large logs during the summer and fall, will likely reveal the fruiting bodies of several different species of myxomycetes (Stephenson and Stempen 1994).

In a forest, fungi may not be as obvious as animals and (especially) plants, but the roles that fungi play are absolutely essential if forest ecosystems are to function properly. It is important that these still understudied and very much underappreciated organisms be given the consideration they certainly warrant (Cripps 2004).

References

Alanbagi RA, Alshuwaili FE, Stephenson SL (2019) Fungi associated with forest floor litter in Northwest Arkansas. Curr Res Environ 9:25–35

Alexopoulos CJ, Mims CW, M. Blackwell M (2007) Introductory mycology, 4th edn. John Wiley and Sons, New York

Alshammari N, Stephenson SL (2018) A preliminary study of wood-decay fungi in forests of northwest Arkansas. Curr Res Environ 8:556–563

Alshammari NI, Veettil VN, Sulieman AME, Stephenson SL (2020) Impact of field and laboratory environmental conditions on the diversity of wood-decay fungi in the forests of northwest Arkansas. J Pure Appl Microbiol 14:1801–1808

Ben Hassine Ben Ali M, Nelsen DJ, Garrett Kluthe B, Collins T, Stephenson SL (2018) Diversity of ectomycorrhizal fungi associated with *Quercus alba* in northwest Arkansas. Curr Res Environ 8:418–424

Bruhn JN, Wettroff JJ, Mihail JD, Kabrick JM, Pickens JM (2000) Distribution of *Armillaria* species in upland Ozark Mountain forests with respect to site, overstory species composition and oak decline. Eur J For Pathol 30:43–60

Coker WC (1923) The clavarias of the United States and Canada. University of North Carolina Press, Chapel Hill

Coker WC, Couch JN (1974) The gasteromycetes of the Eastern United States and Canada. University of North Carolina Press, Chapel Hill

Cripps CL (2004) Fungi in forest ecosystems: systematics, diversity, and ecology. New York Botanical Garden Press, Bronx

Elliott TF, Stephenson SL (2018) Mushrooms of the Southeastern United States. Timber Press, Portland, OR

Elliott TF, Jusino MA, Trappe JM, Lepp H, Ballard G, Bruhl JJ, Vernes K (2019) A global review of the ecological significance of symbiotic associations between birds and fungi. Fungal Divers 98:161–194

Hesler LR (1960) Mushrooms of the Great Smokies: a field guide to some mushrooms and their relatives. University of Tennessee Press, Knoxville

Hyde K (2022) The numbers of fungi. Fungal Divers 114:1. https://doi.org/10.1007/s13225-022-00507-y

Jonsell M, Nordlander G (2004) Host selection patterns in insects breeding in bracket fungi. Ecol Entomol 29:297–705

Kelley MB, Fierke MK, Stephen FM (2009) Identification and distribution of Armillaria species associated with an oak decline event in the Arkansas Ozarks. Forest Pathol 39:397–404

Lee CA, Bruhn J, Muzika R-M (2017) Distribution of *heterobasidion irregulare* in Missouri pine-oak forests. N Am Fungi 12:1–20

Lee C, Holdo R, Muzika R-M (2021) Feedbacks between forest structure and an opportunistic fungal pathogen. J Ecol 12:4092–4102

Miller OK Jr (1977) Mushrooms of North America. E. P. Dutton, New York

Petersen RH (1971) The genera *Gomphus* and *Gloeocantharellus* in North America. Nova Hedwigia 21:1–118

Policelli N, Horton TR, Hudon AT, Patterson TR, Bhatnagar JM (2020) Back to roots: the role of ectomycorrhizal fungi in boreal and temperate forest restoration. Front For Glob Change 3:97. https://doi.org/10.3389/ffgc.2020.00097

Rosen HR (1926) A new *amanita* from Arkansas. Mycologia 18:97–99

Stephenson SL (2010) The kingdom fungi: the biology of mushrooms, molds, and lichens. Timber Press, Portland, OR

Stephenson SL, Stempen H (1994) Myxomycetes: a handbook of slime molds. Timber Press, Portland, OR

Swartz D (1933) Studies of Arkansas fungi. Am Midl Nat 14:714–719

Tedersoo L, Santiage-Ramírez S, Köljaig U, Bahram M, Döring M, Schigel D, May T, Ryberg R, Abarenkov K (2018) High-level classification of the fungi. Fungal Diver 90:135–159

Vasilyeva LN, Stephenson SL (2009) The genus *Diatrype* Fr. (*Ascomycota*, *Diatrypaceae*) in Arkansas and Texas (USA). Mycotaxon 107:307–313

Wargo PM (1977) *Armillaris mellea* and *agrilus bilineatus* and mortality of defoliated oak trees. For Sci 23:435–492

Wheeler Q, Blackwell M (1984) Fungus-insect relationships: perspectives in ecology and evolution. Columbia University Press, New York

Chapter 10
Plants of Special Interest

Abstract The forests and other types of habitats in the Appalachians and Ozarks are characterized by diverse assemblages of plants. Some of these plants are of special interest because of their rarity, use by humans, or some unusual feature. Edible plants, medicinal plants, carnivorous plants, and poisonous plants are the best-known to most people, but there are other plants with some feature that causes them to be unusual in nature, although the particular feature involved may not be readily apparent. The purpose of this chapter is to describe some of these plants of special interest found in the Appalachians and Ozarks. These range from such familiar plants as the common but non-native dandelion to some of the rarest plants in the entire world.

Keywords Edible plants · Medicinal plants · Carnivorous plants · Poisonous plants · Mycoheterotrophic plants · Rare plants

10.1 Introduction

As discussed in some detail in the preceding chapters, forests dominate much of the landscape of the Appalachians and Ozarks. The assemblages of species have changed over time and today vary from place to place. Some localities support highly biodiverse assemblages of plants and other organisms, but in other localities few species are present. Vascular plants such as trees are large enough to be easily noticed, but this is not true of numerous less conspicuous examples, especially many of the so-called lower plants and fungi, although members of these groups are often extraordinarily common and some are of considerable ecological significance. In this chapter emphasis will be placed on vascular plants of special interest for one reason or another. In some instances—but not all—there is an obvious human

---

S. L. Stephenson, R.-M. Muzika, *Forests of the Southern Appalachian Highlands and Interior Highlands of the United States*, Plant and Vegetation 20,
https://doi.org/10.1007/978-3-032-01155-8_10

connection. The plants considered are some of the vernal plants (the "spring wildflowers"), which are also known as spring ephemeals, medicinal plants, a special and distinctive type of wild onion, foods derived from "wild" plants, poisonous plants that can be dangerous, some species of very rare plants that may be in danger of extinction, endemic plants found only in certain areas, and nonnative invasive plants that pose a problem in some portions of the Appalachians or Ozarks.

10.2 Poisonous Plants

Plants that grow in the wild do not come with labels. Our knowledge of what constitutes an edible plant has been acquired over a considerable period. For plants native to the Appalachians and Ozarks, the process began many centuries ago and certainly involved a degree of trial and error. Native Americans passed along some of what they knew to early settlers, who undoubtedly did some trial-and-error testing of their own. Fortunately, many poisonous plants contain alkaloids and other substances that have a bitter taste, which decreases the chances that they will be consumed. This is not always so, and some plants contain carcinogens that lack any trace of a bitter taste. A number of edible plants are easy to recognize and can be consumed by humans, as described above, but there are also some plants that are poisonous. Buckeye (*Aesculus glabra*) produces nuts that are somewhat similar to those of American chestnut (*Castanea dentata*), but they contain a poisonous glycoside that could be fatal if the nuts are consumed by a small child. It is well known that the wilted leaves of black cherry (*Prunus serotina*) are poisonous, but the seeds have high levels of a cyanide-producing glycoside and if consumed by a small child are potentially fatal.

Mayapple (*Podophyllum peltatum*) is a very common and conspicuous plant of moist forests throughout the Appalachians and Ozarks (Gentry et al. 2013). It is easily recognized from its umbrella-like leaves and large white flower. The fruit that develops from the flower is marginally edible, but this is not true of the rest of the plant, which is highly poisonous. Canada moonseed (*Menispermum canadense*), a woody vine most commonly encountered in low, moist forests along streams, is morphologically similar to wild grape and has caused human poisoning. Poison hemlock (*Conium maculatum*) is one of the more famous poisonous plants, and Socrates is thought to have been poisoned by a hemlock-laced drink in 399 BCE. Poison hemlock is native to Europe and the Mediterranean region but now occurs throughout most of North America. The plant itself is not uncommon in moist soils at the edges of ponds and other bodies of water. It is a tall, much-branched herbaceous plant with a smooth, shiny, hollow stem covered with purple spots. All parts of the plant are deadly poisonous.

Among the other poisonous herbaceous plants found in the both the Appalachians and Ozarks are dwarf larkspur (*Delphinium tricorne*), bracken fern (*Pteridium aqulinium*), butterfly weed (*Asclepis tuberosa*), white snakeroot (*Ageratina altissima*), and white baneberry (*Actaea pachypoda*). Dwarf larkspur is the most frequently encountered larkspur east of the Great Plains and is sometimes called "stagger weed" because of the condition the plant produces in cows feeding upon it. All members of the genus *Delphinium* are toxic to both humans and livestock. In the Appalachians, a number of poisonous shrubs can be added to the list. These include mountain laurel (*Kalmia latifolia*), great laurel (*Rhododendron maximum*), and American yew (*Taxus canadensis*). The majority of these are relatively common in the regions in which they occur, but their poisonous properties are of much less concern to humans than to livestock (e.g., when they are incorporated into baled hay).

The poisonous plant most likely to pose problems for humans is poison ivy (*Toxicodendron radicans*), which is most abundant in disturbed forests but also can occur in just about any other type of habitat. This woody vine, with shiny, bright green leaves divided into three leaflets, contains a substance called urushiol, a skin irritant found in all parts of the plant. Many humans are sensitive to the effects of urushiol and develop a skin rash that is often quite painful. Virginia creeper (*Parthenocissus quinquefolia*) closely resembles poison ivy but is not posionous. They both grow in the same habitats and can often be found growing on the same tree. Virginia creeper can be identified by having a leaf divided into five leaflets instead of the three characteristic of poison ivy.

Jack-in-the-pulpit (*Arisaema triphylla*) is one of the more distinctive of all poisonous plants. It belongs to the same family (the Araceae) as skunk cabbage, to be described later in this chapter. Like skunk cabbage, jack-in-the-pulpit produces a fleshy inflorescence (the spadix) that is surrounded by a fleshy leaf-like bract (the spathe). The latter can be completely green in color or green with striking brown stripes. In a given plant, the inflorescence may contain only male flowers, only female flowers, or be bisexual. This is an unusual feature in plants. The tissues of jack-in-the-pulpit contain sharp-pointed crystals of calcium oxalate. When consumed, the crystals irritate the mouth and throat, often causing swelling that can be serious. As such, the nature of the toxicity of jack-in-the-pulpit differs from other plants mentioned above.

10.3 Edible and Medicinal Plants

Although much traditional knowledge has been lost, there are a number of wild plants known to have been of use to Native Americans in traditional medicine. Some of the plants used are poisonous if not administered in proper amounts. The latter is an important point that must be emphasized. Plants used in traditional medicine contain one or more "active ingredients." For the more toxic examples, it is essential that the proper dosage be determined; otherwise, dosages could be lethal. One of the more commonly encountered examples of a medicinal yet poisonous plant in the Appalachians and Ozarks is pokeweed (*Phytolacca americana*), a large herbaceous plant often found in disturbed areas and forest edges. All parts of the plant, especially the root, are poisonous. However, the young leaves are sometimes collected and boiled to remove the poisonous substances, and after the water is discarded the resulting "poke salad" is consumed. The plant has been used to treat many ailments, including acne, tonsillitis, and skin inflammations including boils.

The leaves of various herbaceous plants—and the stems if they were not too tough—were food items for both Native Americans and early settlers. They were especially important as sources of essential vitamins that were often lacking in many of the foods consumed during the winter months. In the spring the basal leaves of such plants as burdock (*Arctium minus*), lamb's quarters (*Chenopodium album*), and yellow rocket (*Barbarea vulgaris*) were collected, cleaned, and then cooked to yield what were often referred to as "spring greens" or "potherbs." All of these grow wild, although yellow rocket (also called wild mustard or winter cress) is planted in gardens. Other plants collected for their underground stems or roots were duck potato (*Sagittaria latifolia*), ground nut (*Apios americana*), and broad-leaved cattail (*Typha latifolia*). Today, only those inclined to try exotic wild foods are likely to consume any of these.

Dandelion (*Taraxacum officinale*) was among the first plants introduced into North America from Europe, an intentional introduction that took place in the early 1600s. Although it is difficult to think of dandelion as a cultivated plant, this had been the case in Europe for centuries. As such, it was only natural that the first colonists brought it along. In Europe, dandelions were a medicinal plant, used as a treatment for various illnesses, including liver problems, gastrointestinal distress, fluid retention, and skin ailments. Moreover, the dandelion was considered to be tasty and highly nutritious. All parts of the plant can be eaten, including the root and flowers. The leaves can be eaten as spring greens, the flowers are used to make dandelion wine, and the boiled roots steeped to make a tea or roasted and used as a substitute for coffee.

The name "dandelion" is a mispronunciation of the French name "*dent de lion*," which translates to "tooth of the lion." This refers to the leaves of the dandelion, which are serrated and look somewhat like teeth. The dandelion plant consists of a rosette of leaves originating from a central growing point. During the growing season, the plants give rise to one to several flower stalks (scapes) 4 to 8 cm long. These produce the unmistakable golden-yellow flowers that ultimately transform into the characteristic spherical, fluffy white seed heads. Each of the latter consists of numerous small fruits (achenes) attached to a parachute-like cluster of fine hairs. The fruits are rather (some would say too much so) effectively dispersed by wind.

Once introduced, the dandelion quickly spread. Native Americans soon discovered the plant and developed their own uses for it. The growing season for dandelions is very long, and an occasional flower can be observed as early as late winter and well past the time most plants are still active in the fall. Dandelions require full sunlight in order to thrive and thus are not inhabitants of closed canopy forests. They are most common in areas subject to some level of disturbance but can be found just about anywhere except where native plants are well established. Dandelion is clearly an example of an introduced plant that has become so naturalized that most people tend to accept it as part of the native flora.

Fig. 10.1 Sassafras (*Sassafras album*) can be easily recognized from the three types of leaves the tree produces. (Photo by Steven L. Stephenson)

10.4 Making Tea

It is possible to make a "wild tea" from many plants, with sassafras tea certainly the best-known. This tea, which has a distinctive flavor, is made using pieces of bark stripped from the roots of a sassafras tree (*Sassafras albidum*), which is relatively common in both the Appalachians and Ozarks (Fig. 10.1). Sassafras is a medium-sized tree that only rarely reaches the forest canopy. It is easily recognized by its trimorphic leaves, meaning that the leaves can be any of three different shapes on the same tree. Some leaves are unlobed, some are two-lobed, and some have three lobes. Sassafras is example of an allelopathic tree (Gant and Clebsch 1975). Allelopathic plants produce biochemicals that influence the survival of other plants. In the case of sassafras, the biochemicals produced have a negative effect on neighboring plants, reducing completion for sassafras. Sassafras tea was once widely consumed by early settlers and also was used in traditional medicine by Native Americans. However, it is now known that sassafras tea is hazardous to humans if consumed regularly.

It has been suggested that it is possible to make a tea from just about anything, and various people have tried teas made from the dried leaves or fruits of such plants as persimmon (*Diospyros virginana*), wild strawberry (*Fragaria virginana*), red clover (*Trifolum pratense*), staghorn sumac (*Rhus typhina*), and mulberry (*Morus rubra*). Several types of wild mints, including bee balm (*Monarda diyma*), are especially good for making tea. Ultimately, what is acceptable is a matter of taste, but most people would consider the tea made from such plants as bee balm quite good.

10.5 Mycoheterotrophic Plants

Among the more unusual plants found in the forests of the Appalachians and Ozarks are those that meet their energy needs in an unusual fashion (Strausbaugh and Core 1978; Gentry et al. 2013). These mycoheterotrophic plants are commonly found in the forest understory where low levels of light limit photosynthesis. The most common and widespread mycoheterotrophic plants are orchids, but mycoheterotrophy also has evolved in other groups of plants. This term is used for those plants that have lost most or all of their chlorophyll during the course of evolution. Since leaves are no longer required to provide an increased surface for photosynthesis, most mycoheterotrophic plants have reduced or no leaves. Because the plant can no longer photosynthesizes, it forms an association with a mycorrhizal fungus in which the plant receives the nutrients needed in order to survive from the mycelium of the fungus. Some orchids do retain their chlorophyll and thus can photosynthesize, but they still need to be mycoheterotrophic in order to be become established after seed germination.

Among the heterotrophic plants found in both the Appalachians and Ozarks are the coralroot orchids of the genus *Corallorhiza*. The name coralroot refers to the coral-like mass from which the stem arises in the soil. Coralroot orchids are small plants, usually only about 8 to 16 cm tall, and easily overlooked since they are not common. The plant has a single erect stem that is generally yellowish, brown, or reddish purple in color. The flowers are generally inconspicuous. Coralroot was used as a treatment for colds, pneumonia, and skin irritations in traditional medicine.

One of the coralroot orchids found in the Appalachians is a very rare plant and serves as an example of how much we do not yet know about the forests of either the Appalachians or Ozarks. Bentley's coralroot (*Corallorhiza bentleyi*) was discovered in 1998 (Bradford 2017). The known distribution of the species is limited to 15 populations located in a small area of Virginia and West Virginia, where it is found in deciduous forests and along roadsides (Fig. 10.2). Bentley's coralroot has a pale red or yellow stem and produces up to 20 small, inconspicuous flowers of the same color.

A larger and better known mycoheterotrophic plant is ghost pipe (*Monotropa uniflora*), named for its translucent "ghostly" white to pale pinkish white color (Fig. 10.3). Ghost pipe ranges in height from 10 to 30 cm. Like the coralroot orchids, ghost pipe has is only a single stem, although the stems of several plants tend to occur together. Each stem bears a single flower, and the leaves are scale-like. When ghost pipe first emerges from the ground, the flower is pendant (downwardly pointed), but as the plant matures, the flower becomes erect (in line with the stem), and the color of the plant ultimately turns to black.

Ghost pipe is common and widespread throughout both the Appalachians and Ozarks, where it can be found most often in forests where oak is present. In traditional medicine ghost pipe was used to treat both physical and emotional pain, but the plant is somewhat toxic and shouldn't be consumed. Because it resembles certain macrofungi, ghost pipe is sometimes mistaken as some type of mushroom (Stephenson 2013).

Fig. 10.2 Bentley's coralroot (*Corallorhiza bentleyi*) is a rare orchid discovered in 1998 and currently known form only a few localities in the Ridge and Valley of Virginia and West Virginia. (Photo by Zach Bradford)

Fig. 10.3 Ghost pipe (*Monotropa uniflora*) is a mycoheterotrophic plant that resembles a fungus more than a plant. (Photo by Kim Fleming)

10.6 A Thermogenic Plant

Spring is a special time of year in the forests of the Appalachians and Ozarks. As first mentioned in Chap. 3, numerous spring wildflowers appear during a short period in the spring (the vernal season), when the forest floor has warmed enough for plant growth to resume but before leaves are fully developed on the trees. On some moist slopes the flowers carpet the forest floor with their colorful display. Virtually all spring wildflowers are perennials and sprout from an underground portion of the plant that overwinters in the soil. A few species produce their flowers very early in the spring, well in advance of most other members of the vernal flora. Skunk cabbage (*Symplocarpus foetidus*) is usually the first harbinger of spring at higher elevations in the Appalachians, sometimes flowering as early as late February (Fig. 10.4). However, the flowers are not at all showy and the plant itself is limited to areas of swampy ground not likely to receive many human visitors so early in the year. However, this plant has one feature not shared with any other plant found in the region—it is thermogenic (Watling et al. 2008). Because skunk cabbage flowers so early in the year, the ground where it occurs is likely still covered with ice or snow. The plant has the capacity to warm up its immediate surrounding by as much as 15 to 30 °C, allowing it to literally melt its way through frozen ground. The heat produced by the plant has a secondary benefit—it helps attract insect pollinators to the skunk cabbage flowers. Skunk cabbage belongs to a family of plants (the Araceae) in which many species are thermogenic. Its name is derived from the unpleasant odor the plant releases.

Fig. 10.4 Skunk cabbage (*Symplocarpus foetidus*) is one of the earliest plants to flower in early spring in the Appalachians. (Photo by Charles Garrett)

10.7 Rare Plants

In any given region there may be a number of endemic species (i.e., species that occur nowhere else). Given the uniqueness of habitats in many areas of the Appalachians and the Ozarks, it is not suprising that certain species of plants are found there and nowhere else. In the Ozarks alone the extent of endemism is estimated at more than 250 species, many of which are plants. The Oachiata Mountains support a disproportional number of endemic species of plants (Zollner et al. 2005). Many endemic species of plants in the Ozarks are associated with glade and riparian habitats; fire-maintained woodlands are also characterized by a number of endemic plants, insects, amphibians, and reptiles. There are numerous examples, some of which are detailed below, but many others are worth mentioning. For example, the Ozarks wild crocus (*Tradescantia longipes*), which is also known as the dwarf spiderwort, is found only in the mountains of southwest Missouri and northwest Arkansas.

Virginia round-leaf birch (*Betula uber*) was once known from only a single population restricted to one small area along the bank of a stream in the Balsam Mountains of southwestern Virginia (Fig. 10.5). This tree, which can grow to a height of about 10 m, has an interesting history. Similar in appearance to black (or sweet) birch (*Betula lenta*) but with distinctive rounded leaves, it was discovered by William Ashe, a botanist from North Carolina, in 1913, but then "lost" to science for more than 60 years. Unfortunately, Ashe mistook where he was when he made his discovery and recorded the wrong locality. As a result, efforts by botanists to find the plant were unsuccessful because they were looking in the wrong place. The tree was finally rediscovered in 1975 (Ogle and Mazzeo 1976). No additional populations of round leaf birch have been found, but the plant has been successfully propagated in botanical gardens and other places. Sadly, this type of successful "resurrection" has not been possible for many other rare plants.

Once the location of the population of roundleaf birch became known, it was subject to being visited by people who wanted to acquire seedings and samples of

Fig. 10.5 When it was first discovered, Virginia roundleaf birch (*Betula uber*) was known from only a single population in the Appalachians. (Photo by Steven L. Stephenson)

the plant. The population of roundleaf birch, which extended over an area of only about 100 m^2, suffered greatly even after the population was enclosed by a fence. Unfortunately, much of the damage was done by research scientists and visits by various groups. Out plantings of the species have been established on the Jefferson and George Washington National Forest, and living specimens are now in arboreta worldwide. One problematic aspect of roundleaf birch still remains. Some experts now recognize the plant as a variety of the common black birch (*Betula lenta*) as *B. lenta* var. *uber*, but the evidence is not yet conclusive to others. In any case, the conservation efforts surrounding roundleaf birch will best serve as an example of the tangled complexity and difficulty involved in preserving a species and will hopefully guide conservation efforts in the future to better outcomes.

The maple-leaf oak (*Quercus acerifolia*) is a similar example of a rare tree and, as expected from its name, has a confounding appearance. The leaves of maple-leaf oak are as wide or wider than they are long, have three to fine prominent lobes, and are similar in shape to the leaves of sugar maple (*Acer saccharum*). The leaves also resemble those of shumard oak (*Quercus shumardii*). Maple-leaf oak can grow to a height of 15 m. The species is known to occur at just four widely scattered sites in the Oauchita Mountains of western central Arkansas. All sites are located in open, rocky woodlands on or near ridgetops. Recently, populations of oak with leaves suggestive of maple-leaf oak have been found elsewhere, but the identification of these has yet to confirmed (Wu et al. 2023). As was the case with the round leaf birch, maple-leaf birch has been successfully propagated in botanical gardens and other places.

Peter's Mountain mallow (*Iliamna corei*) is one of the rarest plants in the world. First discovered in 1927, this plant is known from only a single small population on Peter's Mountain in the Ridge and Valley of southwestern Virginia. Peter's Mountain mallow is a perennial herbaceous plant that can reach a height of a meter and has large (5 cm in diameter) pink flowers (Fig. 10.6). In general appearance it is similar to the ornamental hollyhock (*Alcea rosea*). The plant occurs on a northwest slope near the ridgetop of Peters Mountain, where it grows out of shallow soil on sandstone outcrops in an open forest. The plant does not tolerate shade, and this is only one of several factors that have to be addressed if the species is to survive. The plant was named after Earl Core, a botanist at West Virginia University.

The showy lady's slipper (*Cypripedium reginae*) is perhaps the most beautiful flower found in the Appalachians and Ozarks (Fig. 10.7). It is also one of the rarest flowers in the two region. The species is not rare over most of its range, which encompasses much of the northeastern and north central United States along with southern Canada. The showy lady's slipper is large orchid, reaching a height of more than a meter and producing one to three flowers per stem. The flowers occur at the tips of the stems, with each flower subtended by a leaf-like bract. The leaves on the plant are narrowly ovate, up to 25 cm long, and more than 6 cm wide.

The showy lady's slipper grows in wet habitats such as fens and swamps, usually where the soils are basic to slightly acidic. In the Ozarks it can be found at the base of north-facing limestone bluffs along small streams, in fens, and wet swales. The showy lady's slipper is a prime example of a plant that is subject to having its

Fig. 10.6 Peter's Mountain mallow (*Iliamna corei*) is one of the rarest plants in the world, with the only known population on Peter's Mountain in the Ridge and Valley of southwest Virginia. (Photo by Zach Bradford)

Fig. 10.7 The showy lady's slipper (*Cypripedium reginae*) is perhaps the most beautiful flower found in the Appalachians and Ozarks. (Photo by Charles Garrett)

flowers removed by someone who finds them particularly attractive but does not realize the significance of removing them. The small white ladyslipper orchid (*C. candidum*) is rare and found in only one locality in extreme southern Missouri. Other orchids of note and rariety include early ladies tresses (*Spiranthes lucida*), rare in Missouri, and the rein orchid (*Platanthera flava*), which can reach a height of more than a meter and is known from only five counties in Missouri.

10.8 Carnivorous Plants

Most people find carnivorous (although "insectivorous" is a more appropriate term to use) plants utterly fascinating. Insectivorous plants are uncommon in the Ozarks, at least those examples that occur in terrestrial habitats. In the Appalachians, common sundew (*Drosera rotundifolia*) can be locally abundant in bogs and similar moist sites located at higher elevations but is not encountered elsewhere. In contrast, dwarf sundew (*D. brevifolia*) the more common of the two species of sundew found in the Ozarks, occurs in open grasslands.

Insectivorous plants evolved this unique ability to trap and digest insects in order to survive in places where the soil is poor in nutrients or is so acidic that nutrient availability is severely decreased. They are not common, with only several hundred species known worldwide. The way in which a particular type of insectivorous plant actually captures an insect is not the same for all examples. The leaves of sundews are arranged in a basal rosette. The individual leaves are spoon-shaped, with a long narrow petiole giving rise to a round, highly modified blade on the end. The upper surface of the blade is densely covered with glandular hairs that secrete a sticky substance that forms what looks like a drop of dew at the tip. These glisten in the sun and are thought by some to attract flying insects. Any that insect encounters the hairs becomes entrapped, and its body dissolved by enzymes produced by the plant. Sundews are surprisingly small plants, often on more than 2 to 5 cm across. When they occur in a wet area with abundant *Sphagnum* present, sundews are easily overlooked.

The purple pitcher plant (*Sarracenia purpurea*), which does not occur in the Ozarks, is a much larger plant, with leaves that can reach 30 cm (Fig. 10.8). The leaves are hollow, with a broad opening with an expanded edge at the apex. Their resemblance to a pitcher accounts for their common name. The leaves fill with water when rain occurs. Insects land on the expanded edge of the leaf and venture into the opening, where stiff, downward pointing hairs prevent them from leaving. A few insects eventually fall into the water at the base of the leaf, where digestive enzymes secreted by the plant dissolve the body of the insects. The nutrients released are absorbed by the plant. The water trapped in the leaf of a pitcher plant provides a microhabitat for other organisms. At least two species insects are known use the water in the leaf as a breeding location, and the water itself supports a diverse assemblage of microorganisms. On exceedingly rare occasions, the remains of small mammals, salamanders, and birds have been found in the leaves of pitcher

Fig. 10.8 The purple pitcher plant (*Sarracenia purpurea*) is an insectivorous plant found in the Appalachians but not in the Ozarks. (Photo by Steven L. Stephenson)

plants. Pitcher plants flower in mid-spring, producing a solitary flower at the top of a flower stalk (scape) that reaches well above the leaves. The flower itself resembles a thick, flat disc ringed with dark, maroon-colored petals. The pitcher plant has been purposely introduced to some areas of the Appalachians where it does not naturally occur. One example is Cranberry Glades in West Virginia.

10.9 Other Plants Worth Noting

10.9.1 Sugar Maple

Sugar maple (*Acer saccharum*) is a relatively common tree in the forests of both the Appalachians and Ozarks. When the first Europeans arrived in eastern Americans, they discovered that the Native Americans were using the tree as a source of sugar in the form of maple syrup. In early spring, when the temperatures first arose above freezing, Native Americans would “tap” sugar maple trees in order to collect the sap. This involved carving a shallow trough in the bark and collecting the sap that came out in some type of bark or ceramic vessel. Once filled with the sap, the vessels were left out overnight to freeze, which would separate water from the sugar. The ice was then removed, leaving behind a thick syrup. About 40 liters of sap were needed to produce about a liter of syrup. Dried syrup became sugar that could be stored in cone-shaped bark containers.

Interestingly, sugar maple isn’t the only tree that can be tapped for syrup, since both red maple (*Acer rubrum*) and box elder (*Acer negundo*) also can be used. However, sugar maple produces the largest amount of sap with the highest sugar content. Maple syrup was valuable to Native Americans a number of ways. They used it to cure meats, as a sweetener for bitter substances used in traditional medicine, and as an important trade item. Maple sugar also was an important source of nutrition and energy during the harsh winter months when other food sources were scarce. In the southern part of its range in North America, which includes the southern portion of the Ozarks, sugar maple produces very little sap. As such, the production of maple syrup was more important for Native Americans in the Appalachians than for those in the Ozarks. Since the honey bee was not successfully established in the United States until the mid-1800s, honey would not have been available to early Native Americans. However, many of the wild berries found in the Appalachians and Ozarks have a sweet taste when they are ripe.

10.9.2 Witch Hazel

Witch hazel (*Hamamelis*) is represented by two species in the Appalachians and Ozarks. The common name comes from the use of this plant as a divining rod with the hope of finding sources of water (i.e., as a "witching rod"). Both the American witch-hazel (*H. virginiana*) and the Ozark witch-hazel (*H. vernalis*) occur in the Ozarks, but the American witch-hazel is found only in the Appalachians. Both are small shrubs that only occasionally reach a height of 10 m. Witch hazel flowers have four, strap-like petals. Those of American witch hazel are yellow and those of Ozark witch hazel vary from yellow to brownish red. The petals are rather persistent, lasting much longer than those of other flowers. Moreover, the petals respond to changes in temperature, curling up when the temperature drops on unusually cold days but uncurling on warm days.

The most important difference between the two species of witch hazel relates to when in the year they flower. American witch hazel flowers very late in the year, sometimes as late as early December when there is snow on the ground. In contrast, Ozark witch hazel flowers very early in the year, sometimes by mid-February. In the Ozarks where both species occur, the two witch hazels can produce the last flowers of the season as well as the first flowers of the season. Witch hazel has one other unusual feature. The mature fruits (capsule) split open and eject seeds for distances of 5 to 8 m. Mechanical ejection of seeds is not common, but there are other examples of plants in the Appalachians and Ozarks that possess this same feature. One of the more common examples is spotted touch-me-not (*Impatiens capensis*) in which the fruit springs apart when touched (hence the common name), ejecting seeds in all directions. Another name for touch-me-not is jewel weed.

Witch hazel has a number of medicinal uses that have been known to humans for a very long time. An extract from the plant is an astringent and has been applied to the skin to treat numerous conditions such as eczema and acne. It was used widely before hydrocortisone for inflammation and also to reduce swelling.

10.9.3 Ramps

Several species of wild onions were among the plants consumed by Native Americans in the Appalachians and Ozarks. Meadow garlic (*Allium canadense*) was the most common and widespread example. Of the other edible wild onions found in the Appalachians, one has a certain measure of notoriety. Ramps (*A. tricoccum*) are conspicuous because of their large, erect, bright green leaves and their tendency to be unbelievably abundant on some moist slopes in deciduous forests of the Allegheny Mountains (Fig. 10.9). In some places, ramps can form an almost complete cover over the forest floor. This plant does not flower until early summer. By the time this happens, the leaves have disappeared, and all that remains is an erect flower stalk with a cluster of white flowers at the top and seemingly coming directly out of the ground. Many people who are very familiar with the ramps they collect in the spring have never seen it in flower. Native Americans foraged for ramps, which they seem to have considered a delicacy, and ramps were also collected for the table by early settlers. Ramps gained their notoriety more recently.

Each spring ramp festivals are held each in small towns throughout the mountains of West Virginia, and some people are willing to drive a considerable distance to purchase ramps, sometimes from roadside markets (Core 1966). What is the attraction? Ramps have an overpowering flavor, with an odor to match. When people eat ramps, they breathe and sweat out the odor. This can continue for as long as several days. There is little question that this wild onion, which many consider delicious, has acquired more of a reputation than any other edible plant in the Appalachians.

Fig. 10.9 Ramps (*Allium tricoccum*) are edible wild onions found in moist forests of the Appalachians. (Photo by Steven L. Stephenson)

References

Bradford Z (2017) Corallorhiza in Virginia. Sempervirens. Winter 2017:1–3
Core EL (1966) Vegetation of West Virginia. McClain Printing Company, Parsons, WV
Gant RE, Clebsch EEC (1975) The allelopathic influences of *Sassafras albidum* in old–field succession in Tennessee. Ecology 56:604–615
Gentry JL, Johnson GP, Baker BT, Witsell CT, Ogle JD (2013) Atlas of the vascular plants of Arkansas. The Arkansas Vascular Flora Project, Fayetteville
Ogle DW, Mazzeo PM (1976) *Betula uber*, the Virginia round-leaf birch, rediscovered in southwest Virginia. Castanea 41:248–256
Stephenson SL (2013) A natural history of the Central Appalachians. West Virginia University Press, Morgantown
Strausbaugh PD, Core EL (1978) Flora of West Virginia. Seneca Books, Morgantown, WV
Watling JR, Grant NM, Miller RE, Robinson SA (2008) Mechanisms of thermoregulation in plants. Plant Signal Behav 3:595–597
Wu Y, Hipp AL, Fargo G, Stith N, Ricklefs RE (2023) Improving species delimitation for effective conservation: a case study in the endemic maple-leaf oak (*Quercus acerifolia*). New Phytol 238:1278–1293
Zollner D, MacRoberts MH, MacRoberts BR, Ladd D (2005) Endemic vascular plants of the interior highlands, USA. SIDA Contribut Bot 21(3):1781–1791

Chapter 11
History of Settlement, Land Use, and Logging

Abstract Native Americans have been present in the Appalachians and Ozarks for more than 10,000 years and possibly as long as 14,000 years. Their impact upon the forests was minimal, but the arrival of European settlers in the 1700s ushered in an era in which forests were increasingly disturbed. Forests were first cleared for agriculture and to obtain wood for fuelwood and building homes. Later, systematic logging of large areas of forest occurred, sometimes in conjunction with mining operations. More recently, forests have given way to urban areas, housing developments, roads, rights-of-way, and industrially developed sites. Nevertheless, significant portions of both the Appalachians and Ozarks are forested, albeit with a past history of some degree of disturbance in almost every instance.

Keywords Native Americans · Fuel wood · Logging · Shay locomotive · Coal mining · Sawmills · Early settlers · Fire

11.1 Introduction

As indicated earlier in this book, Native Americans are known to have been present in the Appalachians and Ozarks prior to 10,000 years ago. Their numbers were very low, and their presence had no extensive effects on the forests of the two regions other than leaving an imprint of fire in some localities. These first human inhabitants practiced a hunter-gatherer life style. Available evidence suggests that they hunted mostly small to medium sized animals, both terrestrial and aquatic, with the white-tailed deer, elk, and bison being the most commonly utilized for food, skins, and bone. When Native Americans first arrived during the late Pleistocene, much larger animals (members of the megafauna) were still around. Killing one of these would have presented a major challenge, but there is no question that they were also prey animals.

S. L. Stephenson, R.-M. Muzika, *Forests of the Southern Appalachian Highlands and Interior Highlands of the United States*, Plant and Vegetation 20,
https://doi.org/10.1007/978-3-032-01155-8_11

11.1.1 Food Available from Forests

The forests of the Appalachians and Ozarks contained numerous plants that could provide edible rhizomes, bulbs, seeds, fruits (especially nuts and berries), and other plant parts. Among the nuts were those of black walnut, butternut, hazel-nut, chestnut, beech, oak, and hickory. Fruits included wild grapes, mulberries, crabapples, strawberries, persimmons, papaws, serviceberries, blackberries, raspberries, and blueberries. The bulbs of wild onions and what would later become known to the early colonists as Jerusalem artichoke (*Helianthus tuberosa*) were also part of their diet (Core 1966). Wild herbaceous plants, especially goosefoot (*Chenopodium berlandieri*), were harvested for seeds and leaves, and others were used for medicinal purposes. In addition to these "wild" foods, the Native Americans cultivated corn, beans, and squash (sometimes known as the "three sisters"). Surprisingly, the Native Americans appear to have used relatively few macrofungi (mushrooms) for food. However, some mushrooms also were used for medicinal purposes.

Forests and non-forest habitats (e.g., grassy glades) provide the habitat for a wide variety of animals, as was discussed in Chap. 7. These animals represented an exceedingly important source of food for Native Americans. They probably fed upon any type of animal that could be killed. The main items in their diet were relatively small mammals, including the beaver, rabbit, ground hog, and opossum, that could be killed with a spear or trapped in some fashion. The bow and arrow began to be used about 1000 years ago. Native Americans also would eat snakes, lizards, and turtles along with birds such as the turkeys, geese, and ducks. As already noted, the largest mammal hunted with regularity was the white-tailed deer. Bison, elk, and bear would have been more difficult to kill. In addition to the animals listed above, fish were extremely important items in the diet.

11.1.2 Native Americans and Forests

The first Europeans settlers reached the Appalachians and Ozarks in the 1700s. Native Americans had inhabited the two regions for thousands of years and until the Woodland Period (discussed elsewhere in this book), when agriculture largely replaced hunting and gathering, their forest clearing was not widespread and confined mostly to the rich soil of river bottoms. Early European settlers found evidence of their presence in the form of numerous "Indian fields" abandoned and covered with secondary growth vegetation. Planted patches of hawthorns (*Crataegus* spp.) and remnants of orchards also were commonly encountered.

11.1.3 Arrival of Europeans

Even as late as the 1500s, the presence of Native Americans had a minimal effect on forests. This changed soon after the arrival of the Europeans, whose impact on the forests became increasingly evident as more land was converted to agricultural use and other human activities. However, there are other factors that influenced the co-occupation of Native Americans and European settlers in the Appalachians and Ozarks, especially in the former region, where there were armed conflicts between the two groups. The Ozarks were settled somewhat later than the Appalachians, with the first establishment of settlements occurring as recently as the early 1700s. While the populations of settlers were increasing during the 1600s and 1700s, populations of Native Americans were at their lowest point in centuries.

When the very first Europeans reached North America, they brought with them diseases such as influenza, smallpox, measles, syphilis, and typhus fever. These were well-established in Europe, where people had some immunity, but absent from North America. The indigenous Native Americans had no natural immune defenses for these diseases, and this led to catastrophic epidemics that decimated their populations soon after first contact with European explorers. The extent of the depopulation is not known, but in some places it was estimated to be as high as 75 to 95%. As a result, Native American communities suffered considerable cultural upheaval and societal collapse. The areas occupied by Native American populations contracted drastically. In addition, colonization pressure exerted by Europeans resulted in the displacement and subsequent occupation of Native American lands previously lived on and utilized by them. In some instances, this pressure led to conflict between the two groups. It is no accident that the westernmost settlements in the Appalachians during the mid- to late 1700s were associated with forts.

The first settlers who moved into the southern portion of the Ozarks did so from the east along the Arkansas River Valley. Many were from the states directly east and southeast. This portion of the Ozarks is characterized by relatively level terrain as opposed to the more rugged areas both north and south. Although there were early settlements along the Mississippi River that ultimately gave rise to settlements in the Ozarks, relatively few settlers moved into the northern portion of the Ozarks until after the War of 1812. The settlement patterns were typically along the "bottoms" and a type of linear valley settlement (Schroeder 2002). After the war, the numbers greatly increased. For example, during the period from 1814 to 1820, the population of Missouri, much of which falls within the Ozarks, swelled from 16,000 to 70,000. By 1860 the population of Missouri was 1.2 million. The rugged terrain of the northern Ozarks made movement from place to place difficult, and the settlements that were established tended to develop close kinship ties not unlike those associated with the settlers who lived in the mountains of the Appalachians.

11.2 Forests as a Source of Wood

Throughout their history, humans have used fuelwood for fires to cook food and for heating shelters or dwellings. This was true for both indigenous Native Americans and early settlers in the Appalachians and Ozarks. Typically, the fuelwood was obtained from the surrounding forest. These uses, coupled with the need for building materials, began to make extensive inroads into the amount of forest cover. When coal became readily available in the United States after the Civil War, it replaced fuelwood in many homes. The first settlers in the two regions usually constructed simple dwellings (cabins) using logs from the surrounding forest and rocks from the immediate area. These usually log-constructed cabins were typically small with limited floor space and often designed for both living quarters and defense in the early years of European settlement. What a colonial era dwelling would have looked like can been seen in the replica of a late-1700s example that is part of the Fort Witten exhibit located near Tazewell, Virginia (Fig. 11.1). Note should be made of the overhanging second story to allow the occupants of the dwelling to use weapons to fire upon the enemy.

The first dwellings were constructed without the use of little more than an axe. Wood saws were not yet available. Most early dwellings had a dirt floor. The number of trees that went into the construction of a dwelling was not great, although their removal created openings in the forest. Additional trees were needed for the construction of any outbuildings such as small sheds and for building "split-rail fences" to enclose livestock. An example of a "split-rail" fence in the Pea Ridge National Military Park in northwestern Arkansas is shown in (Fig. 11.2).

In the late 1700s and 1800s, commerce in much of the northern Ozarks was based on the fur trade, first between Native Americans and the early settlers and then between hunters and fur companies. The fur companies established some of the earliest European settlements in this region of the United States. The animals involved in the fur trade included the white-tailed deer, raccoon, black bear, beaver, otter, and mink, of which the beaver was most important. Although not located in the Ozark region, the Missouri Fur Company in St. Louis was among the major fur trading companies.

Fig. 11.1 Replica of a late-1700s dwelling that is part of the Fort Witten exhibit located near Tazewell, Virginia 1770s house. (Photo by Steven L. Stephenson)

Fig. 11.2 Split rail fence at the Pea Ridge National Historical Park in the Ozarks of northern Arkansas. (Photo by Steven L. Stephenson)

11.3 Other Products from the Appalachians and Ozarks

The export of lead and salt also was important. Given the geology, the Ozark region is an important source of lead, zinc, and copper in addition to other minerals. Some of the largest deposits of lead in the world are found in the Missouri Ozarks. Lead has been mined in the upper St. Francis River region of what is now Missouri since the 1720s in what is now known as the "old lead belt." Over time, lead mining expanded toward southwest Missouri and then into Arkansas by the early 1800s. The importance of lead as well zinc and other minerals and their smelting is often neglected in the history of Ozark region. Soil contamination, dangerous air quality, implications for human health, and damage to ecological resources (including forests) have been well documented. Much of the old mining sites are currently under active restoration, including superfund sites.

Most early settlers largely based their livelihood on livestock, mostly pigs, rather than crops other than corn. Pigs were often allowed to run free in forests, allowing them to feed upon acorns and various plants. Their feeding activities (rooting) undoubtedly disturbed the understory vegetation and reduced the numbers or eliminated the plants they favored in some localities. Anecdotally, "rooting" is often implicated as a cause for the absence of regeneration of certain desired tree species. Hunting remained important until later in the 1800s. Early settlers in both the Appalachians and Ozarks invariably grew corn (or maize), as the Native Americans had done in the same regions before them. The so-called Indian corn grown by early Native Americans encompassed numerous different genetic varieties and was more colorful than today's more genetically uniform corn. In any case, it was necessary to grind (or mill) the corn into a flour with which to bake cornbread for their meals. The Native Americans used simple grinding stones for this, as sometimes did the early settlers. However, the grinding stones used by settlers were sometimes much more elaborate. During his travels in Missouri in 1834, Featherstonhaugh (1844) encountered hand-operated sweep-mills and, in one instance, described the operation of this method of grinding corn.

Fig. 11.3 Historic Mabry Mill in the southern Blue Ridge of Virginia. (Photo by James Mitchell)

> They had no mill of any sort to go to, but had scooped out a cavity in the stump of a large tree, over which was a wooden pestle, eight feet long, suspended from a curved pole 16 feet [5.0 m] in length, with a heavy weight at the end of it. A cross stick was fixed in the pestle, about two feet [0.6 m] from its base; so putting the grains of maize into the cavity, and laying hold of the cross stick, we pounded away…

An "expedition" by Henry Rowe Schoolcraft across the Ozarks of Missouri and Arkansas during 1818–1819 provided much information about cultural details of the early Ozarks way of life. He traversed a broad circle from Bonne Terre across the central Ozarks, as far south as what is now Batesville, Arkansas, and back along eastern Ozarks (Raffery 1996).

It could be argued that the technological advance that had the greatest impact on early settlers was the appearance of the waterwheel-driven flour mill (Fig. 11.3). These mills had to be located near a stream. However, the land along streams usually had the most fertile soils, which thus attracted settlers. Because flour was essential for baking, the miller was an important person in the communities where a flour mill existed. Large springs in the Ozarks represented important gathering locations; southern Missouri and northern Arkansas represent one of the highest concentrations of freshwater springs in North America. Large mills were found throughout, with some of the most notable examples occurring along Big Spring, Greer Spring, Mammoth Spring, Alley Spring, and Blue Spring.

11.4 Saws, Sawmills, and Logging

The first saws to be used in both the Appalachians and Ozarks were crude, hand-operated saws (also called whip saws or pit saws) consisting of a narrow blade with a handle on each end. To operate the saw, one man would stand on the top of a felled log and the other worked out of a pit dug into the ground. This was a rather inefficient and very labor intensive method of sawing timber, but it did allow large logs to be reduced to planks and beams. As a result, the dwellings constructed from these could be referred to as proper houses and not just simple dwellings.

Manual cutting was soon replaced by water-powered or steam-powered saws. After the invention of the circular saw in 1777 and its rapid implementation in the early 1800s, sawmills began to appear, first in the Appalachians and then in the Ozarks. This was accompanied by a major increase in the amount of logging carried out and a resultant reduction in the areas covered by forests. Efficient logging required that the trees logged could be transported to a sawmill and the lumber produced transported to where it could be used or sold. This was not possible until two technological advances were made in the late 1880s, as described below.

From 1861 to 1865, some portions of the Appalachians and Ozarks suffered greatly from the turmoil and violence of the Civil War. One of the major theaters of the war was the Shenandoah Valley. As a result of military operations, numerous farms were destroyed, pillaged, or abandoned, large numbers of livestock killed, and large areas of forest cleared or burned (Stephenson 2013), to provide support for the many thousands of soldiers from both sides of the conflict who were moving and camping throughout the area. The loss of these forests is probably the least appreciated consequence of the Civil War. One report published in the late 1870s indicated that forests in the mountainous areas west of the Shenandoah Valley were reduced by more than a third between 1860 and 1870. Nevertheless, large tracts of forest still existed.

It has been estimated that 4 million hectares of uncut forest remained in just West Virginia in 1870, with much of it represented by red spruce forests located at higher elevations on the Allegheny Plateau. Amazingly, these forests were almost completely logged by 1920, just 50 years later. Two technological advances made such a dramatic rate of logging possible. The first was the introduction of a faster, more efficient type of saw to cut logs into boards (Clarkson 1964). This new type of saw, called a bandsaw, used a long, thin, continuous band of steel blade, usually 8–12 inches wide and with teeth along one side. The blade was mounted between two large wheels. Early examples were water-powered, which required them to be located along a stream. However, during the late 1800s, bandsaws began to be powered by steam and it was no longer necessary for them to be located near a stream. The second technological advance was the development of a special geared locomotive designed by a man with the name Ephraim Shay. This locomotive, which was to bear the name of its inventor (Fig. 11.4), had features that allowed it to operate on the steep mountain slopes and sharp curves that could not be negotiated by conventional locomotives.

Logging during this period involved felling the trees, cutting the logs into sections, pulling them out of the forest with teams of horses or mules (Fig. 11.5), loading them on flatcars, and then using the Shay locomotive to haul them to a sawmill lower on the mountain. After arriving at the sawmill (Fig. 11.6), the sections of logs would be cut into boards. During the late 1800s and early 1900s, the large-scale logging carried out in the Appalachians led to the development of spike and logging camps and "boom" towns throughout the forests. Logging and spike camps were temporary but located so as to allow relatively easy access for the lumber men who

Fig. 11.4 Still operational Shay locomotive of the Cass Scenic Railroad in the Allegheny Mountains of West Virginia. (Photo by Steven L. Stephenson)

Fig. 11.5 Pulling logs out of the forests of the Appalachians with teams of horses or mules in the late 1800s or early 1900s. (Photo from American Lumberman, 1910)

Fig. 11.6 Early 1900s sawmill in the Allegheny Mountains of West Virginia. (Photo from Forest and Wood Industries, 1910)

Table 11.1 Amount of lumber produced by a sawmill in Davis, West Virginia

Type of lumber	Board feet	Cubic meters
Red spruce	439,517,452	1,037,146
Hemlock	272,450,227	642,911
Various hardwoods	450,180,040	1,062,307

This is for the period of 1887 to 1924. (Adapted from Clarkson 1993)

worked in the areas being actively logged. These varied in size, from a spike camp where work would be completed in a few days to a typical camp where residence might be several months in duration. These camps were made up of perhaps 50 to 60 men, each of whom had a particular job to do. Housing was primitive at best, but meals were supplied. The typical routine involved working from dawn until dusk 6 days a week. Wages were low, but logging supplied jobs for thousands of men during its most active period. This often resulted in the migration of lumber men from the Northeast and Canada, who were brought in by rail to work until the timber had all been shipped out. Boom towns developed around the large sawmills which received the logs from trees that were being cut. These had houses where the logger's families could live along with such things as stores, churches, a saloon, and oftentimes the Appalachian equivalent of a brothel. Most of these boom towns survive today only as names on old maps because the loggers often returned to their original homes. The best and most detailed account of all aspects of logging in the mountains of West Virginia is provided in Roy Clarkson's *Tumult on the Mountains* (Clarkson 1964). An indication of the amount of wood harvested from forests of the Appalachians during the late 1800s and early 1900s is the amount of lumber produced by a sawmill in Davis, West Virginia (Table 11.1).

The logging history of the Ozarks is rich and well chronicled. As early as 1818 there was logging activity in the upper Gasconade River in Missouri, where forests of shortleaf pine dominated. By 1880s the best timber in the Gasconade has been removed. The forests of Ouachita Mountains, which occupy the southernmost portion of the region, remained largely untouched until almost the end of 1800s, with only a few sawmills operating in the region during the period of the 1870s to about 1900. These sawmills were small, temporary, and could be moved from one place to another in response to the availability of timber to cut. Prior to the appearance of these sawmills, farmers in the region lived in log-type cabins; raised cows, hogs, and chickens; and grew corn, field peas, and cotton. Lumber from the sawmills allowed them to construct more modern-appearing "box" houses. The pre-logging forests of the Ouachita Mountains consisted mostly of shortleaf pine (*Pinus echinata*), with an admixture of various species of hardwoods. Smith (1986) described these forests in his book *Sawmill: The Story of Cutting the Last Great Virgin Forest East of the Rockies* as truly magnificent. He indicated that it was once said that one of the individual pine trees could yield enough lumber to build a three-room house!

Fig. 11.7 Locomotive pulling flatcars loaded with logs in the Ouachita Mountains of southern Arkansas in the early 1900s. (Photo courtesy of Roberts Library of Arkansas History and Art)

The logging era for the forests of the Ozarks boomed between 1880 and 1920. In Arkansas specifically, it began with the formation of the Crossett Lumber Company in 1899 and the subsequent establishment of large permanently located sawmills in 1908 (Darling and Bragg 2008). By then, large sawmills were putting out 150,000 board feet of timber per day, with most of it being pine. However, the same sawmill also could produce as much as 50,000 board feet of hardwood timber per day. (A board foot is the equivalent of a single board 12 inches by 12 inches by 1 inch, with 1 inch equal to 2.54 cm.) "Sawmill towns" developed at the sites of the larger sawmills, a situation not unlike what happened in the Appalachians. In the early 1900s, the Grandin mill in Missouri represented one of the largest in the world. The Grandin operation consumed 70 acres of woodland a day and produced an average of 60 million board feet a year board feet per day (Cunningham and Hauser 1989). Once again, shortleaf pine was the species of interest in logging operations.

At first, the same methods used in the Appalachians were applied to logging in the Ouachita Mountains. Logs were dragged or hauled out of the forest with teams of horses or skidded down a slope, ultimately to be loaded on flat cars, and the flat cars pulled to the sawmill with a locomotive (Fig. 11.7). The very first locomotives used were Shay locomotives, but they were soon discarded for conventional locomotives. The Shay locomotive was simply too slow (no more than 24 km/h), and the less rugged terrain didn't require the level of maneuverability of this type of locomotive. Logging lasted in the Ouachita Mountains until about 1950, when the supply of available timber was essentially exhausted. As such, large-scale logging lasted longer in the Ozark region than in the Appalachians. In the latter region, few extensive tracts of forest remained after about 1920. Although shortleaf pine was extensive and valuable, it was exhausted throughout much of the Ozarks by 1920. Oak, particularly white oak, was very valuable for use as railroad ties throughout the logging era. For most of the 20th century and currently, Ozark white oak is priced for making barrels.

11.5 Fire

11.5.1 Native Americans and Fire

Fire is a natural component of all ecosystems, and the probability, frequency, and severity of fire differ dramatically across the globe and within forest types. It is not possible to generalize fire across the whole of the Ozarks or across the entire Appalachians since many factors must be considered. An unintended consequence of extensive logging was fire. This is particularly true in the Great Lakes region where logging was followed by the largest fires that ever occurred in North America (until the present century!). It was also common that fires occurred in logging debris in the Ozarks.

Fire is an environmental factor of considerable importance in forests. Before the arrival of Native Americans in the Appalachians and Ozarks, virtually all fires were caused by lightening. It is likely that throughout much of the regions, they were comparatively rare and affected only a small percentage of the total forest cover. However, the frequency and ecological impact of fire undoubtedly increased with the arrival of Native Americans. There is compelling evidence that wherever there were Native Americans, there were fires. Fire history reconstructions reveal that changing human population densities (both indigenous and colonizers)correspond with fire activity. Studies have revealed a strong association of fire with the occurrence of indigenous populations who purposely used fire for a number of reasons, including driving game from cover, clearing the forests for agriculture, and reducing the undergrowth to enhance the growth of certain plants such as blueberries and to make it easier to gather acorns and chestnuts (Tice 1987). In general, studies revealed that fires caused by settler activity was more intensive, extensive and frequent than fires attributed to indigenous or natural ignitions.

11.5.2 Early Settlers

The early settlers viewed forests as a hinderance, since they needed open land to build their settlements and grow their crops. Setting fire to an area of forest in order to clear the land was not an unusual occurrence. As human activities increased, the incidence of accidental fires increased. Brooks (1911) reported on what he encountered in one area of the Allegheny Plateau in West Virginia, indicating that.

> The havoc wrought by forest fires in this section is calculated to fill a lover of woods with dismay. The country about the heads of Gandy Creek, Seneca Creek, and Big Run has been burned over so often that much of the land is cleared of all timber. There are thousands of acres (1000 acres = 405 ha) on which there remains scarcely a trace of vegetation except a rank growth of bracken fern (*Pteridium aquilinum*) and club moss (*Lycopodium* spp.).

Fires were often associated with logging, and the logging of the high-elevation red spruce forests of the Allegheny Plateau represents one example that might be considered as a worst case scenario. The forests being logged were characterized by a thick layer of plant litter/humus on the forest floor. This litter/humus layer usually remained moist under the shelter provided by the canopy of the red spruce trees, but when the trees were logged, there was no canopy, and the litter/humus layer dried out. This made it subject to being burned, and this often happened for several reasons, some intentional and some accidental. The tree laps were often just piled up beside the locomotive tracks, and when dry they were simply a tinderbox. Sparks from the stack or firebox of the wood-fueled Shay locomotives represented a frequent ignition source. Once the dry litter/humus/tree laps were ignited, they burned away almost completely, leaving behind mostly bare rocks that sometimes cracked from the intense heat. Succession under such conditions proceeds very slowly, and some of the burned sites have yet to be recolonized by vegetation well over a century after the fires occurred. It is ironic that some of these usually high elevation sites are now among the most prized and highly valued recreational areas because of the expansive views afforded by lack of forests. Dolly Sods, Grayson Highlands State Park, and the high country of the Mount Rogers Recreation Area are good examples of this phenomenon in the Appalachians, while the Lodge at Mount Magazine represents one located in the Ozarks.

In the Ozarks, there was a distinct association of fire with both indigenous settlements and European settlers. For example, in the Missouri Ozarks fire frequency and occurrence from 1650 to the early 1800s were modest but slowly increased with immigration from Native Americans from the east and increasing numbers of European settlers. Fire occurrence was greatest from 1850 to approximately 1890. After that period of time, agricultural development became more important, Moreover, burning over the previous decades had depleted the fuel load (Guyette et al. 2002). The pattern was similar in the Arkansas Ozarks. However, the episodes occurred more recently since settlement was later in the Arkansas Ozarks. From a study along the Buffalo River Wilderness, Stambaugh and Guyette (2006) determined that fires occurred every 7.7 years during the 1600s and 1700s. Fires increased around 1820 to such an extent that there was a fire every 2 years until 1920. Moreover, fire history data confirm that topographic conditions moderate or promote fires. For example, fire in an area with steep hills is both less frequent and less severe than fire in a woodland, glade, or prairie type of landscape. There is a strong association of fire with drought conditions but only prior to European settlement. However, when looking at human activity and population density, it is clear that humans can override the effects of topography and drought. Simply put, humans represent the primary source of fire (outside of naturally caused lightning fires), and as human populations increased, fires followed.

11.5.3 Prescribed Burning

Prescribed burning (or controlled burning) has become a common management technique for forests. In most respects, prescribed burning may differ little from what Native Americans were doing thousands of years ago, although today (as would be expected) it is more scientifically based and conducted for specific objectives (Ryan et al. 2013). There are countless reasons that prescribed burning has become a more common practice. For example, it has long thought to be a way to promote wildfire habitat. For forestry practices specifically, low intensive surface fire is used to reduce competition for desired regenerating tree species such as oak. Fire is also used to promote pine regeneration and to encourage diversity of herbaceous plants, forbs, and other species that likewise promote insect diversity. As such, fire is an important tool for ecological restoration. More recently, there has been an emphasis on prescribed burning to reduce the amount of coarse woody debris so as to prevent catastrophic fires. Other reasons for prescribed burning include reducing the spread of insect pests and tree diseases, removing unwanted (usually invasive) plants that threatened native plants, and enhancing the growth of plants fed upon by wildlife. Moreover, the burning of living and dead components of a forest ecosystem recycles nutrients back into the soil.

The timing of prescribed burning is critical, and specific conditions are necessary both to allow the fire to become established while also minimizing and controlling the spread and reducing its potential danger. Precise conditions for such factors as temperature, humidity, wind speed and wind direction, and moisture of the vegetation will determine both the success and control of the fire. Smoke is also a major factor that has consequences for human health. Even when burn plans have been outlined to consider all of these things, conditions on the ground on the day of the proposed burn may cause it not to take place on a given day. For any type of prescribed burning, maintaining control of the fire itself is the biggest concern, and the vast majority of fires are low-intensity surface fires that remain well below the level of the tree canopy.

Prescribed burning has always been a controversial subject. The charring left on the bark of trees and the reduction of the litter layer are both readily apparent. Moreover, prescribed burning, especially when done on a large scale, can generate a considerable amount of smoke. Sometimes, this smoke can be observed many kilometers away from the site being burned. Nevertheless, prescribed burning has proven to be an effective tool for forest management and continues to be used. In a typical year approximately 38,000 km^2 in the United States is subjected to prescribed burning. Prescribed burning is also used in non-forested habitats such as glades, grasslands, and very often in prairies, so this total would reflect that fact. It is possible that future climate change may affect prescribed burning by decreasing the number of days when meteorological conditions are suitable for burning. Moreover, there are some aspects of prescribed burning that probably require further investigation. One of these relates the high temperatures generated on the forest floor when the litter layer burns. Presumably, this could affect soil microbes, but there are little data on this aspect of prescribed burning. In addition, when coarse woody debris burns, the associated wood-decay fungi are killed. Once again, not enough is known to evaluate this situation.

11.6 Coal Mining

As indicated earlier in this book, the layers of sedimentary rocks that underlie the Allegheny Plateau and the Cumberland Plateau have interbedded coal seams (Fig. 11.8). At first, when there was not yet a major demand, the coal used came from seams exposed at or very near the surface of the ground. However, with the rapid expansion of the industry in the years following the Civil War, the demand for coal greatly increased. This ultimately required the use of extensive underground mining, sometimes with a mine shaft extending for a considerable distance underground. Working a mine required manpower, and small towns developed in areas where there was active mining. The presence of these small towns and everything needed to support the mining operation affected nearby forests, which were cut or severely degraded. However, the coal mines were a major factor in the total economy of the states (primarily Kentucky, Virginia, and West Virginia in the region of Appalachia considered in this book) in which they occurred. However, when coal seams are located close to the surface, they can be exposed by removing the overlying soil and rock (the so-called overburden). This process, referred to as strip mining or surface mining, has been practiced for a century in places where there is coal. Many of the areas of the Appalachians where strip mining was carried out were forested, and the forests in these areas were totally destroyed. In the 1960s, a special type of strip mining (called mountaintop removal) began to be practiced. Mountaintop removal is what the name implies. It involves removing a major portion of the summit or main ridge of a mountain to permit easier access to coal seams (Fig. 11.9). In areas that are forested, the trees present are cut and either burned or taken away. After the coal has been excavated, the overburden is either replaced to its approximate original location or dumped into an adjacent valley (what is called a valley fill). There is no question that strip mining in all of its forms is detrimental to forests. Fortunately, mining for other minerals such as magnesium, limestone, and iron ore has decreased dramatically within the Appalachians.

Fig. 11.8 Exposed coal seam on the Allegheny Plateau in northern West Virginia. (Photo by Randy Darrah)

Fig. 11.9 Example of mountaintop removal in the Appalachians. (Photo by Vivian Stockman, www.ohvec.org)

Our knowledge of the environmental effects of strip mining are far advanced than they were in the past when no reclamation was carried out in areas that had been surface mined, and these areas were often left barren and subject to erosion (Brooks 1966). Today, these mined areas are reshaped to approximately the original landscape contour. There are several possible options for reclamation of these mined areas. First, they can be converted to managed pasture for grazing wildlife. Pasture can be developed on the previously mined area in a relatively short period of time. Second, the area can be converted into a commercial forest plot. Obviously, any trees established on a plot require years to grow to sufficient size to be harvested. However, in the meantime, the trees sequester carbon, which is of environmental significance. Third, if the location is suitable, the mined area can be developed for outdoor recreation. Regardless of the option pursued, the resultant cover of vegetation is aesthetically more pleasing than an open surface mine.

Historically, most mine operators have chosen the pasture option (Gairola et al. 2023). The primary species of plants seeded on these sites include KY-31 tall fescue (*Festuca arundinacea*), perennial ryegrass (*Lolium perenne*), birdsfoot trefoil (*Lotus corniculatus*), and sericea lespedeza (*Lespedeza cuneata*). These species are used because they are inexpensive, although some are invasive, the seeds are readily available, and the plants can be established quickly. This combination has the added advantage because it includes two members of the family Fabaceae and thus are nitrogen-fixing plants. More contemporary restoration efforts involve an understanding of ecological site conditions and appropriate planting and habitat restoration (Adams 2017).

11.7 Reforestation and Land Use

When reforestation efforts occur, the "new" forest is much less diverse and often contains species that were not part of the original forest (Stephenson 2013). Another aspect of this return to primary succession is the likelihood of invasive species introduction and expansion. Invasive species are among the biggest problems in many forests today.

The greatest change in land use in the Appalachians and Ozarks since 1900 has been the retirement of farm land, which reached its peak during the 1920s and 1930s and continues today. The amount of land currently in crops is probably no more than half that of 1900. Land in pasture has increased but still occupies a small portion of the landscape. Developed land, including urban areas, housing developments, roads, rights-of-way, and industrially developed sites, have increased. Surprisingly, the collective effect of these changes has been an increase in forest cover. However, most of this increase can be attributed to the development of successional forests on disturbed sites, and forest composition is usually quite different from the forests that originally occupied those sites.

Important changes in land ownership have taken place. Increases in absentee ownership by industrial mining and forest corporations, along with increases in government ownership, have been consistent trends in the two regions. One type of land use that has greatly increased since 1900 is the development of natural lands for recreational resorts and second homes. Unfortunately, both invariably involve the removal of forests.

One other aspect of the removal of forests through logging or for other reasons should be mentioned. As a result of the deforestation that has taken place over the past two centuries, erosion and floods have become more prevalent in the Appalachians and the Ozarks. For example, during and immediately following logging, before the cover of vegetation can be reestablished through succession, water is no longer retained as it was when the forest was still intact. During peak agricultural use, early in the 1900s, massive erosion occurred when large low pressure storms hit the mountains. In addition to erosion, the increased flow of water downslope caused streams and rivers to overflow their banks, often with serious consequences for streamside roads, farms, and towns.

11.8 Conservation

Fortunately, concerns for conservation increased alongside the expansion of industry and other human activities into forests. A greater appreciation of the forest emerged as more people began to recognize the value of maintaining forest ecosystems and their assemblages of associated organisms. By the early 1900s, both public and private agencies were involved in managing forests and the landscapes they occupy. This led to the establishment of parks, state and federal forests, wilderness and national recreation areas, and natural area preserves. These have provided refuges for native plants and animals and help to ensure their survival. Conservation efforts have fluctuated throughout late 1900s and early 2000s in response to economic trends, historical events, and politics. Pressure on the forest resources of the Appalachians and Ozarks remains intense as different groups seek to use these important natural resources in many different ways.

References

Adams MB (2017) The forestry reclamation approach: guide to successful reforestation of mined lands, General Technical Report NRS-169. U.S. Department of Agriculture, Forest Service, Northern Research Station, Newtown Square, PA. https://doi.org/10.2737/NRS-GTR-169

Brooks AB (1911) Forestry and Wood Industries, West Virginia geological survey, vol 5. ACME Publishing Company, Morgantown, WV

Brooks DB (1966) Strip mine reclamation and economic analysis. Nat Resour J 1:13–44

Clarkson RB (1964) Tumult on the mountains. McClain Printing Company, Parsons, WV

Clarkson RB (1993) Destruction of the upland forests by lumbering and fire. In: Stephenson SL (ed) Upland Forests of West Virginia. McClain Printing Company, Parsons, WV, pp 35–46

Core EL (1966) Vegetation of West Virginia. McClain Printing Company, Parsons, WV

Cunningham B, Hauser C (1989) The decline of the Missouri Ozark forest between 1880 and 1920. In: Waldrop TA (ed) Proceedings of Pine-Hardwood mixtures: a symposium on the management and ecology of the type; 1989 April 18–19; Atlanta, Georgia, General Technical Report E-58. U.S. Department of Agriculture, Forest Service, Southeastern Forest Experiment Station, Asheville, NC, pp 34–37

Darling OH, Bragg DC (2008) The early mills, railroads, and logging camps of the Crossett Lumber Company. Arkansas Hist Quart 67:107–140

Featherstonhaugh GW (1844) Excursion through the Slave States, from Washington on the Potomac, to the Frontier of Mexico, with Sketches of Poplar Manners and Geological Notices. Harper and Brothers, New York

Gairola SU, Bahuguna R, Bhatt SS (2023) Native plant species: a tool for restoration of mined lands. J Soil Sci Plant Nut 23:1438–1448

Guyette RP, Muzika R-M, Dey DC (2002) Dynamics of an anthropogenic fire regime. Ecosystems 5:472–486

Raffery MD (1996) Rude Pursuits and Rugged Peaks: Schoolcraft's Ozark Journal, 1818–1819. University of Arkansas Press, Fayetteville

Ryan KC, Knapp EE, Varner JM (2013) Prescribed fire in north American forests and woodlands: history, current practice, and challenges. Front Ecol Environ 11(1):e15–e24. https://doi.org/10.1890/120329

Schroeder WA (2002) Opening the Ozarks: A Historical Geography of Missouri's Ste. Genevieve District, 1760–1830. University of Missouri Press, Columbia

Smith KL (1986) Sawmill: the story of cutting the Last Great Virgin Forest East of the Rockies. University of Arkansas Press, Fayetteville

Stambaugh MC, Guyette RP (2006) Fire regime of an Ozark wilderness area, Arkansas. Am Midl Nat 156:237–251

Stephenson SL (2013) A natural history of the Central Appalachians. West Virginia University Press, Morgantown

Tice DA (1987) A history of Albemarle's forests. Mag Albemarle Cty Hist 45:19–60

Chapter 12
Areas of Forests Within the Appalachians and Ozarks

Abstract Most of the forests of the Appalachians and Ozarks were logged prior to the 1920s, but some escaped logging. Today, both regions are largely dominated by forests, and these forests are afforded varying degrees of protection. Forests are almost totally protected in wilderness areas, national parks, and some natural areas. Other forests such as those in state parks, state forests, and some types of limited access areas receive considerable protection from at least human intervention, but most other forested areas are potentially subject to varying degrees of disturbance. The types of disturbance are dependent on regulations and utilization, and the latter includes the direct impact of certain human activities, including logging. Experimental forests were set up as forest laboratories where the effects of various types of disturbance could be studied. There is no question that forests represent a renewable and sustainable resource of incalculable aesthetic, environmental, and economic value. The protection of forests and the habitats with which they are associated remains critical, and doing this is a real challenge because of the wide variation in ownership, ranging from public protection and management to private landowner.

Keywords Wilderness areas · Experimental forests · Forested areas · National forests · National parks · Natural areas · State parks · Protection

12.1 Introduction

Most of the forests within the Appalachians and Ozarks were logged prior to the 1920s, although some areas of forest escaped logging. The most extensive of the latter are found within the Southern Blue Ridge of the Great Smoky Mountains National Park of eastern Tennessee and western North Carolina. However, the more common situation is for non-logged areas to be relatively small. An example is the Gaudineer Scenic Area in the Appalachian Mountains of West Virginia, where a surveying error allowed about 55 ha of red spruce (*Picea rubens*) to survive logging

S. L. Stephenson, R.-M. Muzika, *Forests of the Southern Appalachian Highlands and Interior Highlands of the United States*, Plant and Vegetation 20,
https://doi.org/10.1007/978-3-032-01155-8_12

operations that were otherwise complete in the general area (Core 1966). In places in the Ozarks, many of the best "old growth" forests are associated with springs and mills since those have been gathering areas for the local communities for well over a century. It is not unusual for scattered old-growth trees to be found on cliff faces in the Appalachians and Ozarks, with red cedar (*Juniperus virginiana*) the most likely tree to be encountered in remote, inaccessible areas.

Because of the successional changes that took place after logging, the composition of the forests that currently exist in the two regions certainly differs to some extent from what it was prior to logging. In some instances, other factors such as the chestnut blight fungus (*Cryphonectria parasitica*) in the Appalachians have contributed to significant changes in composition for some forests. In much of the Ozarks, an alteration of the natural disturbance regime such as fire suppression resulted in notable changes in forests. In some high-elevation red spruce forests of the Appalachians, removal of red spruce trees by logging resulted in changes in the microclimate for the forest floor that favored hardwood trees over red spruce. Consequently, there was a complete replacement of one forest type with another. Similarly in the Ozarks, the removal of vast pine forests gave way to more oak-dominated forests within the last 100 years.

Although logging permanently removed forests in some areas of the Appalachians and Ozarks, especially where settlements developed, large portions of both regions are still forested. Moreover, some areas that were forested and later converted to agriculture have reverted back to forests as numerous small farms were abandoned as a result of shifts in population that took place in the 1930s and 1940s. Today, the best place to observe relatively intact areas of forests are the national forests and national parks that occur in the Appalachians and Ozarks (Miller 2016; Waterman 2019). The former are the George Washington and Jefferson National Forests, Monongahela National Forest, Wayne National Forest, Pisgah National Forest, Daniel Boone National Forest, Cherokee National Forest, Nantahala National Forest, Mark Twain National Forest, William B. Bankhead National Forest, Chattahoochee National Forest, Sumter National Forest, Ozark National Forest, and Ouachita National Forest. Only three national parks (Great Smoky Mountains and Shenandoah in the Appalachians and Hot Springs in the Ozarks) are located within the two regions. A vast area of the Ozarks includes lands managed by the U.S Park Service. The Ozark National Scenic Riverways in Missouri was established in 1964 as the first national park to protect a river system. This area includes the Current and the Jacks Fork Rivers. In addition, the Buffalo National River in Arkansas is protected by the U.S. Park Service and was established in 1972. Protections of these rivers provide extensive additional protection of a vast area of associated forests as well. The karst geology and topography of the Ozarks results in clear, spring fed rivers, the largest springs in North America. It is worth noting that there are several state parks and other areas under state jurisdiction that serve to protect forested landscapes.

12.2 National Forests

The George Washington National Forest in west central Virginia and the Jefferson National Forest in southwest Virginia were administratively combined in 1995 to form the George Washington and Jefferson National Forests. The two national forests encompass 6734 km^2 in Virginia, 500 km^2 in West Virginia, and slightly less than 4 km^2 in Kentucky. This national forest extends over portions of the Blue Ridge, Ridge and Valley, Allegheny Plateau, and Cumberland Plateau physiographic provinces. Elevations range from 157 to 1743 m on the summit of Mount Rogers. The George Washington and Jefferson National Forests contain 23 designated wilderness areas, the largest of which is the Mountain Lake Wilderness Area in the Ridge and Valley province of southwest Virginia with a small portion in adjacent West Virginia. Other wilderness areas include Peters Mountain, Beartown, Brush Mountain, Ramseys Draft, and Rough Mountain. Because of their vast extent and the wide range of elevations at which they occur, considerable variation exists for the forests of the George Washington and Jefferson National Forest. Forests with at least an admixture of pines often occur on more xeric sites, but the predominant forests are mixed oak. Subalpine coniferous forests, including one example with Fraser fir (*Abies fraseri*) present, are found at the very highest elevations. Descriptions of representative portions of the George Washington and Jefferson National Forests were provided by Adams and Stephenson (1983, 1991), Stephenson et al. (1993), and Stephenson (2013).

The Monongahela National Forest in West Virginia covers 3726 km^2 and extends over a portion of the Allegheny Plateau physiographic province (including the Allegheny Mountains on the eastern edge of the province). Elevations range from 270 to 1482 m on the summit of Spruce Knob. The Monongahela National Forest contains eight designated wilderness areas, the most notable of which are Cranberry, Dolly Sods, Laurel Fork North, Laurel Fork South, and Otter Creek. The Spruce Knob-Seneca Rocks National Recreation Area (covering an area of 400 km^2) contains the highest point (Spruce Knob) in the region along with Seneca Rocks, a large quartzite rock outcrop. Many of the forests of the Monongahela National Forest would be considered as mixed mesophytic, but extensive areas of northern hardwood forests also exist. Higher elevations support alpine coniferous forests, with balsam fir (*Abies balsamea*) present at a few localities. Forests of the Monongahela National Forest were described by Core (1966), Stephenson and Clovis (1983), Bryant et al. (1993), and Muzika et al. (1999).

The Wayne National Forest encompasses 989 km^2 of the Allegheny Plateau physiographic province in southeastern Ohio, which is at the northwestern edge of the region considered as the Appalachians in this book. The highest elevation is 349 m on Barber Ridge. There are no designated wilderness areas in the Wayne National Forest. This reflects the history of the land now occupied by the national forest. During the 1700s and 1800s, all of the forested land was logged and then used for agriculture. Coal mining also had a major impact during the 1800s and 1900s. As such, the forests currently present largely represent the result of a reforestation project. These forests are mostly mixed hardwoods, with oak and hickory often the most important trees present. Areas dominated by pine and eastern red cedar are not uncommon.

The Pisgah National Forest is located in the Southern Blue Ridge of western North Carolina. This forest covers 2075 km^2 and encompasses several mountain summits that exceed 1800 m. Elevations range from 235 to 2019 m. There are two predominant forest types present over most of the national forest. Cove forests containing such trees as tulip poplar (*Liridodendron tulipifera)*, basswood (*Tilia americana*), white ash (*Fraxinus americana*), cucumber tree (*Magnolia acuminata*), black birch (*Betula lenta*), and black cherry (*Prunus serotina*) occur on the mesic sites, while forests containing various mixtures of oak, hickory, and pine are found on more xeric sites. The main species of oak are white (*Quercus alba*), northern red (*Q. rubra*), and chestnut (*Q. montana*). Spruce-fir forests occupy the very highest elevations (U.S. Department of Agriculture 2023).There are three designated wilderness areas. These are Linville Gorge, Middle Prong, and Shining Rock. Interestingly, the gametophytes of several types of predominantly tropical ferns that reproduce independently of their sporophytes have been reported from the surfaces of shady, moist rocks in such places as Linville Gorge (Farrar 1967).

The Daniel Boone National Forest is located on the Cumberland Plateau of eastern Kentucky and encompasses an area of 7080 km^2. Elevations range from 145 to 1245 m on the summit of Big Sag Ridge. There are two designated wilderness areas. These are Clifty and Beaver Creek. Other sites of special interest include Cumberland Falls, Red River Gorge, and Yahoo Arch. Forests throughout this portion of the Cumberland Plateau consist mainly of such trees as chestnut oak, red maple (*Acer rubrum*), black oak (*Quercus velutina*), American beech (*Fagus grandifolia*), yellow poplar, and sugar maple (A. *saccharum*) (Muller 1982).

The Cherokee National Forest occupies a small portion of western North Carolina, but most of it occurs in eastern Tennessee along the border with North Carolina. The Cherokee National Forest has two separate sections—a northern region to the northeast of the Great Smoky Mountains National Park and a southern section to the southwest of the park. This national forest covers 2653 km^2 in the Blue Ridge physiographic province. Elevations range from approximately 365 to 1915 m on the summit of Roan High Knob. The forests are similar to those of the Great Smoky Mountains National Park. The Cherokee National Forest contains 11 areas that have been designated as wilderness. These include Gee Creek, Pond Mountain, Bald River Gorge, Big Laurel Branch, and Unaka Mountain.

The Nantahala National Forest covers 2150 km^2 in the Blue Ridge physiographic province of western North Carolina. Almost 24 km^2 of this (in the Joyce Kilmer-Slickrock Wilderness Area) is old growth forest. Elevations range from 366 to 1768 on the summit of Lone Bald. Forests in the Nantahala National Forest are similar to those described for the Pisgah National Forests (U.S. Department of Agriculture 2023). There are two wilderness areas other than the Joyce Kilmer-Slickrock Wilderness Area. These are Ellicott Rock and South Nantahala. Portions of the Nantahala National Forest are noteworthy because they are characterized by the highest annual totals of precipitation of any region of eastern North America.

The Mark Twain National Forest consists of nine non-contiguous units in the Ozark Plateau physiographic province of the southern half of Missouri The forest covers a total area of 12,419 km^2, but only about half of this is publicly owned. Elevations range from 75 to 519 m on the summit of Bell Mountain and within the vast area there are 500 caves and among the oldest rocks in North America. The Mark Twain National Forest contains seven designated wilderness areas for a total of >31,000 ha. These include Devils Backbone, Hercules Glades, Bell Mountain, Irish Wilderness, Rockpile Mountain, Paddy Creek, and Piney Creek along with the Eleven Point Wild and Scenic River. Each wilderness area has its distinct features that include a variety of natural communities such as glades, fens, woodlands, riparian forests, old growth shortleaf pine, old growth mixed oaks. As with much of the Missouri Ozarks, the forest was nearly completed harvested in the early 1900s with the focus on harvesting pine for lumber and oaks for staves and ties. Now, across the Mark Twain National Forest more than 100,000 acres (>40,000 ha) are under active restoration using prescribed fire to create a mix of pine forests, pine-oak forests, pine woodlands, and glade habitat. Fire is also used to restore oak dominated forests and oak woodlands.

The William B. Bankhead National Forest covers 733 km^2 on the Cumberland Plateau in northwestern Alabama. Elevations range from 145 to 345 m on the summit of Eckerberger Mountain. The William B. Bankhead National Forest has a single designated wilderness area, the Sipsey. However, this wilderness area is quite large, extending over slightly more than 100 km^2. Information on the William B. Bankhead National Forest was provided by Schweitzer et al. (2015).

The Chattahoochee National Forest occupies 3039 km^2 in the Blue Ridge physiographic province of northern Georgia. The highest elevation is 1458 m on Brasstown Bald. The Chattahoochee National Forest contains ten designated wilderness areas. Among these are the Tray Mark Trail, Brasstown, and Blood Mountain. Four of the wildness areas (Big Frog, Cohutta, Ellicott, and South Nantahala) extend beyond the Chattahoochee National Forest into two other national forests.

One portion (an administrative unit) of the Sumter National Forest in South Carolina in the extreme northwestern part of the state falls within the Blue Ridge physiographic province. This administrative unit of the national forest is the Andrew Pickens Ranger District, which encompasses approximately 344 km^2. The highest elevation is 1002 m on the summit of Fork Mountain. Forests of the Andrew Pickens Ranger District consist of a mixture of shortleaf pine and various hardwoods on low ridges and south-facing slopes. Table Mountain pine (*Pinus pungens*) and pitch pine (*P. rigida*) occur on higher ridges, while oak-hickory forests occupy lower and north-facing slopes (U.S. Department of Agriculture 2004). The Andrew Pickens Ranger District administrative unit includes one designated wilderness area. This is Ellicott Rock, which extends beyond the Sumter National Forest to include parts of the Nantahala National Forest in North Carolina and the Chattahoochee National Forest in Georgia.

The Ozark National Forest covers 4856 km^2, with most of it located on the Ozark Plateau in northwestern Arkansas. The southern portion of the National Forest extends along the Arkansas River Valley south to the Ouachita Mountains. Elevations range from about 225 to 748 m on Turner Ward Knob. There are five designated wilderness areas. These are East Fork, Hurricane Creek, Leatherwood, Richland Creek, and the Upper Buffalo. Oak-hickory forests are most common, but shortleaf pine (*Pinus echinata*) forests occur in some places. Dogwood, red maple, and eastern red bud (*Cercis canadensis*) are consistently present, and American beech occurs in some unusually mesic situations. Descriptions of the forests of the Boston Mountain portion of the Ozark National Forest were provided by Stephenson et al. (2007).

The Ouachita National Forest occupies 7221 km^2 in the Ouachita Mountains physiographic province in the western part of Arkansas and a small portion of southeastern Oklahoma. The highest elevation is 816 m on the summit of Rich Mountain. The forest contains six designated wilderness areas. These are Black Fork Mountain, Dry Creek, Upper Kiamichi River, Flatside, Poteau Mountain, and Caney Creek. The two major forest types on the Ouachita National Forest are oak-hickory and oak-pine. Shortleaf pine-loblolly pine (*Pinus taeda*) forests are not uncommon, and in a very few places there are forests with trees (e.g., beech and red maple) characteristic of more mesic sites. Information on the composition of the forests of the Ouachita National Forest was provided by the U.S. Department of Agriculture (1999, 2005).

Wilderness areas within national forests are protected once they have been so designated, but logging (often extensive) frequently occurred prior to this happening. For example, this was the case for the Otter Creek Wilderness in the Monongahela National Forest in West Virginia. Selective logging in a national forest has always taken place, and a recent executive order from the president of the United States opened up more than half of all national forests to logging.

12.3 National Parks

Relatively few areas of the Appalachians and Ozarks have been set aside as national parks, but one of these (the Great Smoky Mountains National Park) is the best known and most visited national park in eastern North America. The other national parks in the two regions are Shenandoah National Park in the Blue Ridge portion of the Appalachians and Hot Springs National Park in the Ozarks. However, the watershed of the Buffalo National River in Arkansas is managed by the National Park Service, as are the Current and Jack's Fork Rivers in the Ozarks. These two rivers make up the Ozark National Scenic Riverways, created by an Act of Congress in 1964. This was the first river system to be recognized by the National Park Service.

The Great Smoky Mountains National Park occupies an area of 2114 km^2 in western North Carolina and eastern Tennessee (Fig. 1.3 in Chap. 1). The Great Smoky Mountains, for which the park is named, are a subrange of the Southern Blue Ridge Mountains. Sixteen of the mountains located within the park have summits that exceed 1800 m, with the tallest (Clingmans Dome) reaching 2025 m. The forests of Great Smoky Mountains National Park include all of the types found in the Appalachian region, including high-elevation coniferous forests, mixed mesophytic forests, northern hardwood forests, mixed oak forests, beech forests, and xeric pine-dominated forests. Tree diversity in some of the mountain valleys (coves)

Fig. 12.1 Forests on the western slope of Shenandoah National Park from near summit of Hawksbill Mountain, the highest point in the park. (Photo by Gary P. Fleming)

is as high as anywhere in the temperate zone of the world. Moreover, there are extensive areas of old-growth forest. Some of the more detailed descriptions of the forest vegetation of the park were provided by Whittaker (1956), White et al. (1993), and MacKenzie and White (1998).

Shenandoah National Park (Fig. 12.1) encompasses 789 km^2 of the northern Blue Ridge physiographic province in Virginia Elevations range from 163 to 1235 m on the summit of Hawksbill Mountain. A road named the Skyline Drive extends for 169 km along or just below the ridgetop of the entire length of the park, providing easy access to high-elevation forests. Almost 40% of the park (322 km^2) has been designated as wilderness areas. Mixed oak forests are predominant throughout the park, but mixed mesophytic forests occur in sheltered positions on north-facing slopes. There are two small areas with species characteristic of subalpine coniferous forests, and forests with species of pine as major components exist in a number of the more xeric situations. Detailed descriptions of the forest vegetation were provided by Young et al. (2006) and Fleming (2012).

Hot Springs in the Ouachita Mountains of Arkansas was set aside as the Hot Springs Reservation by an Act of Congress in 1832 and designated as a National Park in 1921. If the Hot Springs Reservation is recognized as an early equivalent of a national park, then it would predate Yellowstone National Park, which was created by an Act of Congress in 1872. The land encompassed by the current boundary of the park was set aside as a protected area in 1832, long before the modern concept of national parks even existed. The park encompasses natural hot springs, which is the reason it was afforded protection so long ago. The area of land originally set side in 1832 has greatly increased in extent to its present size of 22.5 km^2. The park is unusual in that it includes a portion of the city of Hot Springs.

The actual hot springs flow from the western slope of Hot Springs Mountain (Fig. 12.2). The immediate area around the springs has been subjected to considerable commercial development.

However, forests are present on the mountain slopes surrounding the springs. These forests are dominated mostly by species of oak and hickory, but dominance shifts to shortleaf pine (*Pinus echinata*) on ridgetops and other xeric sites.

The Buffalo National River (Fig. 12.3) in the Ozark physiographic province of northern Arkansas was designated by an Act of Congress as the first National River

Fig. 12.2 Hot Springs Mountain in Hot Springs National Park. (Photo by James St. John)

Fig. 12.3 The Buffalo National River in the Ozarks of Northern Arkansas. (Photo by Tony Collins)

in 1972. The river originates in the Boston Mountains and flows across the Springfield Plateau and Salem Plateau. The total length of the river is 246 km, but only the lower 217 km is managed by the National Park Service. Three segments of the river and their adjacent watersheds (a total area of about 48 km^2) have been designated as the Buffalo River National Wilderness. The forests of the watershed of the Buffalo National River are mostly mixed hardwoods, but mixed mesophytic forests with such species as beech present occur in some sheltered situations (Stephenson et al. 2007). The Ozark Scenic Riverways is under the management of the National Park Service and encompasses 134 miles of the free flowing, spring fed Jack's Fork and Current Rivers; it also includes the largest spring in North America.

12.4 State Parks

Numerous state parks exist throughout the Appalachians and Ozarks. Many of these have been created for a variety of reasons related to the preservation of sites and/or structures of cultural or historical interest. Others have been created largely to serve as places where recreational activities can be carried out. These activities include such things as hiking, horseback riding, fishing, swimming, and camping. Some state parks include cabins that can be rented out for short periods of time.

It is not surprising that some state parks, especially those in which hiking or riding trails exist, contain areas of forest. In some instances, the areas of forest can be relatively little disturbed (at least recently) and thus likely similar in composition to the original forests of the region in which the state park is located. This is especially true for some parks located at higher elevations in the Appalachians. Two well-known examples are Blackwater Falls State Park and Canaan Valley Resort State Park in the Allegheny Mountains of West Virginia. Canaan Valley Resort State Park also provides the special habitats required for a number of plants that are either rare or of special interest. State parks in the Ozarks include areas with vast forests and unique natural areas such as glades, woodlands, and caves and include the highest elevation in Missouri, in the St Francois mountains, and in Arkansas as Magazine Mountain State Park.

12.5 State Forests

State forests are forested areas administered by the Department of Forestry, Department of Natural Resources, Department of Conservation, or Forest Commission of a particular state. These forests are not nearly as well-known as national parks or state parks, although they sometimes cover a considerable area. For example, there are seven state forests located in the western portion of the state of North Carolina that fall within the Appalachian region as delimited in this book. These range in size from less than 1.0 km^2 to more than 40 km^2. Some of these are not open to the public whereas others require a permit to enter or are open to the public. Some state forests consist largely of tree plantations, while a few such as Paul State Forest in the Shenandoah Valley of Virginia contain a mature, relatively undisturbed forest. Where permitted, state forests are often major centers for hunting, especially for white-tailed deer.

12.6 Natural Areas

In both the Appalachians and Ozarks as well as elsewhere in the United States, there has been a recent movement towards designating specific areas of the landscape as natural areas with the objective of protecting examples of particular habitats or populations of locally, nationally, or globally rare plants or animals. These protected areas vary considerably with respect to their management (Povilitis 2002) and the type of land upon which they occur. Most are on land owned by the state in which they are located, but others are on private land, on federally owned land such as in national forests, on university properties, or on land owned by conservation organizations such as The Nature Conservancy.

One of the best known natural areas in the Appalachians is the Cranberry Glades Natural Area in the Allegheny Mountains of West Virginia. Cranberry Glades is a

Fig. 12.4 The Pinnacle Natural Area Preserve in extreme western Virginia encompasses examples of most of landforms and forest types found in the southern Ridge and Valley. (Photo by Gary P. Fleming)

large bog containing a number of plants characteristically found at more northern latitudes. The Pinnacle Natural Area Preserve (Fig. 12.4) is not nearly as well-known as Cranberry Glades but serves the same purpose of affording protection to an ecologically special place. This natural area is located along the Clinch River in the Ridge and Valley physiographic province of the extreme western part of Virginia. The Pinnacle Natural Area encompasses 464 ha and consists of a landscape made up of relatively steep slopes, mountain ridges, and a stream valley. Except for the pinnacle-like cliff that gives this natural area its name, the Pinnacle Natural Area Preserve is forested. The forests and woodlands present include oak-hickory forests, alluvial forests, and cove forests as well as areas dominated by such species as hemlock and eastern white cedar. Plant diversity in this natural area is unusually high, and the forests, although second-growth, have not been logged recently. Already relatively undisturbed, they ultimately should develop into fully mature forests. The Pinnacle Natural Area Preserve serves as an excellent example of the topography and forests characteristic of the southern portion of the Ridge and Valley. This natural area is open to limited public access. The Chesney Prairie Natural Area in the Ozarks of northern Arkansas is one of the few tall grass prairie remnants on the Sringfield Plateau.

The Devil's Knob-Devil's Backbone Natural Area (Fig. 12.5) is located in the Ozarks of northern Arkansas. The main topographic feature is an uplifted ridge with two knolls joined by a saddle called the Devil's Backbone. This natural area has forests (mixed hardwood and cedar glade woodland), three types of glades (limestone, dolomite, and sandstone), and rocky bluffs that support several species of rare plants. The range of habitats in the natural area encompasses most of those known from the Ozarks. Ashe's juniper (*Juniperus ashei*) is represented in this natural area by a number of very old trees. The Devil's Knob-Devil's Backbone Natural Area is open to public access.

Missouri alone has 183 Natural Areas in property of various ownerships from federal, state, local, and private. Nearly two thirds of these fall within the Ozarks. Arkansas has 78 designated natural areas and 28 of these fall within the Ozarks and Ouachitas regions. These area represent unique habitats that may have once been

Fig. 12.5 The Devil's Knob-Devil's Backbone Natural Area in the Ozark Mountains of Arkansas. (Photo courtesy of the Arkansas Natural Heritage Commission)

more abundant or host endemic species or species of conservation concern, or there is uniqueness in the plant community, the geology, or any combination of those.

Some natural areas allow free access by the public, whereas others require special permission to enter. A major consideration has always been the management of natural areas. Natural areas, as is the case for any other part of the total landscape, are subject to successional change as well as the inadvertent invasion of non-native plants or animals. Moreover, environmental stress in the form of an unusually severe weather event can adversely affect any of the organisms present. Unfortunately, humans also pose a constant threat.

12.7 Experimental Forests

Experimental Forests are forested areas established for research purposes and administered by the U.S. Departure of Agriculture Forest Service (Adams et al. 2008). They are literally "forest laboratories" used for testing types of silvicultural practices, forest manipulations, or various other field-based experiments related to forestry or forest ecology. There are two experimental forests in the Appalachians. These are the Fernow Experimental Forest in the Allegheny Mountains of West Virginia and the Coweeta Hydrologic Laboratory in the Blue Ridge physiographic province in Georgia. The Fernow consists of 1860 ha of forest surrounded by the Monongahela National Forest, and the Coweeta consists of 1818 ha of forest surrounded by the Nantahala National Forest. Four experimental forests occur in the Ozarks. One of these (the Alum Creek Experimental Forest, which covers 1885 ha) is located in the Ouachita physiographic province, while three others—the Henry R. Koen Experimental Forest (290 ha), the Sylamore Experimental Forest (4290 ha), and the Sinkin Experimental Forest (1666 ha)—are located on the Ozark Plateaus physiographic province. Experimental forests are not protected per se except for the fact the areas they encompass remain largely forested.

Research carried out on experimental forests to some extent reflects the types of forest they contain. For example, for those with oak forests, research topics relate to oak regeneration tend to be important. Some experimental forests such as the Fernow contain a series of watersheds in which streamflow, water quality, and forest hydrology can be monitored or examined. These watersheds, which range in size from less than a hectare to several hectares, provide invaluable data on the factors that are necessary to consider when attempting to maintain watershed integrity. The data these smaller examples provide can be extrapolated to larger natural watersheds that may extend over many km^2. For example, a study to evaluate the effects of the pesticide Dimilin®, used to control the spongy moth on non-target organisms was carried out on one watershed in the Fernow in the early 1990s (Gottschalk and Fosbroke 1995).

In the Sinkin experimental forest on the Mark Twain National forest, research has long focused on silvicultural practices and developing guidelines for managing Ozark pine and oak forests. Recently, there has been an emphasis on ecological restoration, oak regeneration, oak decline, and unevenaged forest management.

12.8 Other Forested Areas

In addition to the forested areas mentioned above, some National Historical Sites, Conservation Areas, National Wildlife Refuges, and National Military Parks can have appreciable areas of forest. For example, 1500 ha of forests, some containing individual trees more than 200 years old, occur on the Pea Ridge National Military Park in the Ozarks of northwestern Arkansas. Areas of forest also exist on private property throughout the Appalachians and Ozarks, and some of these exceed more than just a few hectares. Many land owners will grant access, with permission, to their property. This often happens for neighbors who wish to hunt on the property or (especially in the spring for morels) look for mushrooms.

12.9 Degrees of Protection

Protection of forests—both conceptually and practically—spans an immense range. Forests have always been afforded various degrees of protection. In the past, especially in areas where coal mining took place or any lands subjected to surface activity such as development, there was no protection, and large areas of land were completely cleared. Total removal of tree biomass and usually major disruption of the soil (or the entire landscape in strip mining) has serious consequences for reestablishment of forest communities through natural succession. In such instances, reforestation, often with non-native species, has proven to be more effective in providing a forest canopy. In contrast, forests are almost totally protected in wilderness areas, national parks, and natural areas. Forests in state parks and some types of limited access areas receive considerable protection, but most other areas of forest are subject to varying degrees of disturbance. The types of disturbance are dependent on regulations, utilization and include the direct impact of certain human activities (e.g., use of off-road vehicles) to removal of trees by logging. Some forests are now being logged environmentally with horses to minimize the impact. To be completely removed is the ultimate fate of trees in plantation forests, where trees are grown specifically for the purpose of being logged. At least this leaves behind soil suitable for reestablishment of a forest. This was frequently not the case for earlier logging of natural forests, which often destroyed the soil after trees were removed.

Sometimes, the protection afforded a forest can be lifted for economic reasons, with one example being the construction of gas pipelines. The United States has a natural gas pipeline system that carries natural gas nationwide. Across the entire country, there are more than 200 pipeline systems and they extend over a distance of more than 490,848 km. When constructing a pipeline, the most direct route tends to be the most efficient when cost is considered. In many instances, this has meant constructing a pipeline through forested areas. This involves clearing a corridor through the forest for the equipment required to bury the pipe in the ground. Sometimes, construction of a pipeline has been controversial, with the ecological costs versus the economic benefits being hotly debated. One recent example is the Mountain Valley Pipeline, which is a 42 inch (106 cm) pipeline that extends 490 km from northern West Virginia to south central Virginia. This route takes the pipeline across portions of the Allegheny Plateau, Ridge and Valley, and Blue Ridge. A portion (5.6 km) of the pipeline is in national forests.

There is no question of the need to protect forests when such is possible, and this is especially true for relatively undisturbed forests on private property. Moreover, the protection afforded to any forest should extend to maintaining the ecological integrity of the entire watershed in which the forest occurs. Otherwise, the protection is not complete.

One recent regional effort to protect forests is the 500-Year Forest Foundation. The foundation was formed to conserve mature forests and their associated organisms, with the ultimate objective of potentially allowing them to achieve old-growth status. The 500-Year Forest Foundation, which was formed in 1996, is based in Virginia, and not all of the forests involved in this effort are within the Appalachian region. The forests being conserved are on private land, generally at least 40 ha in size, contain a significant number of trees at least 70 years in age, are relatively undisturbed, and are likely to remain largely undisturbed. These forests are protected by conservation easements that should continue indefinitely. The foundation recognizes that areas of true old-growth forests are uncommon or do not occur over most of the eastern United States but are exceedingly valuable from a biodiversity standpoint as well as their ecological significance (e.g., as carbon sinks). Individual land owners also have the option of placing areas into land trusts, a type of easement that can ensure that forested areas can be conserved and managed in perpetuity.

In summary, forests represent a renewable and sustainable resource of incalculable value and one that must be protected without question. The most obvious value to humans is their economic value. Forests can be indirectly utilized in many different ways. For example, the wood from trees provides lumber for construction of many types of structures or objects, including musical instruments, fuelwood (a considerable amount of which is processed for use and is now being exported, mostly to northern Europe), and the bark of trees as a source of mulch. Even more important is their ecological value. The forest canopy provides shade for cooling water in streams and ponds, nutrients for understory herbaceous and shrubby plants, and a breeding site (e.g., hollow logs) for small animals. The type of forest canopy, evergreen or deciduous, largely determines the species composition of everything that lives underneath it. The overall forest habitat is essential for wildlife, including animals for sport hunting and fishing, and as a place for recreation activities such as hiking. New psychological studies have shown both mental and physical benefits are derived from spending time with trees. In addition to recreational activities, a forest is subject to bio-harvesting for various plants and for the simple pleasure of getting out, relaxing, and hunting for nuts, berries, other fruits, and mushrooms.

References

Adams HS, Stephenson SL (1983) A description of the vegetation on the south slopes of peters mountain, southwestern Virginia. Bull Torrey Bot Club 110:18–22

Adams HS, Stephenson SL (1991) High–elevation coniferous forests in Virginia. Va J Sci 42:391–399

Adams MB, Loughry LH, Plaugher LL (2008) Experimental forests and ranges of the USDA Forest Service, General Technical Report NE-321. USDA Forest Service, Northern Research Station, Newtown Square, PA

Bryant WS, McComb WC, Fralish JS (1993) Oak-hickory forests (western mesophytic/Oak-Hickory Forests). In: Martin WH, Boyce SC, Echternacht AC (eds) Biodiversity of the Southeastern United States: upland terrestrial communities. John Wiley and Sons, New York, pp 143–201

Core EL (1966) Vegetation of West Virginia. McClain Printing Company, Parsons, WV

Farrar DR (1967) Gametophytes of four tropical fern genera reproducing independently of their sporophytes in the southern Appalachians. Science 155:1266–1267

Fleming GP (2012) The nature of the Virginia flora. In: Weakley AS, Ludwig JC, Townsend JF, Crowder B (eds) Foundation of the Flora of Virginia Project Inc., Richmond. Botanical Research Institute of Texas Press, Fort Worth, pp 24–75

Gottschalk KW, Fosbroke SL (1995) Proceedings of the 10th Central Hardwood Forest Conference, Morgantown, West Virginia, General Technical Report NE-197. U.S. Department of Agriculture, Forest Service, Northeastern Forest Experiment Station, Radnor, PA

MacKenzie MD, White PS (1998) Vegetation of great smoky mountains national park. Castanea 63:323–336

Miller C (2016) America's great national forests, wildernesses, and grasslands. Rizzoli International Publications, New York

Muller RN (1982) Vegetation patterns in the mixed mesophytic forest of eastern Kentucky. Ecology 63:1901–1917

Muzika R-M, Stephenson SL, Adams HS, Lawrence DW, Miller GW (1999) Patterns of composition in the tree stratum of the fernow experimental forest and adjacent portions of the otter creek wilderness area. In: Eckerlin RP (ed) Proceedings of the Appalachian Biogeography Symposium, Virginia Museum of Natural History Special Publication No. 7. Virginia Museum of Natural History, pp 34–44

Povilitis T (2002) What is a natural area. Nat Areas J 21:70–74
Schweitzer C, Lemke D, Tadesse W, Wang Y (2015) Assessment of forest management influences on total live aboveground tree biomass in William B. Bankhead National Forest, Alabama. In: Holly A, Conner G, Haywood KF, Deds J (eds) Proceedings of the 17th Biennial Southern Silvicultural Research Conference e-General Technical Report SRS-203. U.S. Department of Agriculture, Forest Service, Southern Research Station, Ashville, NC
Stephenson SL (2013) A natural history of the Central Appalachians. West Virginia University Press, Morgantown, WV
Stephenson SL, Clovis JL (1983) Spruce forests of the Allegheny mountains in central west Virginia. Castanea 48:1–12
Stephenson SL, Ash AN, Stauffer DF (1993) Appalachian oak forests. In: Martin WH, Boyce SC, Echternacht AC (eds) Biodiversity of the Southeastern United States: upland terrestrial communities. John Wiley and Sons, New York, pp 255–303
Stephenson SL, Adams HS, Huebner CD (2007) Upland forest vegetation of the Ozark mountains in northwestern Arkansas. Rhodora 109:197–221
U.S. Department of Agriculture, Forest Service (1999) Ozark-Ouachita highlands assessment: terrestrial vegetation and wildlife, Report 5 of 5. General Technical Report SRS-35. U.S. Department of Agriculture, Forest Service, Southern Research Station, Asheville, NC
U.S. Department of Agriculture, Forest Service (2004) Revised land and resource management plan: Sumter National Forest, Management Bulletin R8-MB 126A. U.S. Department of Agriculture, Forest Service, Southern Region, Atlanta, GA
U.S. Department of Agriculture, Forest Service (2005) Revised land and resource management plan: Ouachita National Forest Arkansas and Oklahoma, Management Bulletin R8-MB 124A. U.S. Department of Agriculture, Forest Service, Southern Research Station, Asheville, NC
U.S. Department of Agriculture, Forest Service (2023) Nantahala and Pisgah National Forests: final land management plan Rb-MB 160. U.S. Department of Agriculture, Forest Service, Southern Research Station, Asheville, NC
Waterman J (2019) National Geographic Atlas of the National Parks, 3rd edn. National Geographic, Washington, DC
White PS, Buckner ER, Pittillo JD, Cogbill CV (1993) High-elevation forests: northern hardwoods forests and associated communities. In: Martin WH, Boyce SC, Echternacht AC (eds) Biodiversity of the Southeastern United States: upland terrestrial communities. John Wiley and Sons, New York, pp 305–337
Whittaker RH (1956) Vegetation of the Great Smoky Mountains. Ecol Monogr 26:1–80
Young JA, Foster JER, Townsend PAA (2006) Vegetation of Shenandoah National Park in relation to environmental gradients, U.S. Department of the Interior, National Park Service Technical Report. National Park Service, Philadelphia

Chapter 13
Threats to Forests

Abstract Forests in the Appalachians and Ozarks are subject to a variety of threats. These threats can involve the forests as a whole or just certain species or groups of tree species or other plants. Many of these threats emerge as a result of the introduction, inadvertent or intentional, of plants, insects, or fungi not native to the two regions. Among the most noted examples was the introduction of the chestnut blight fungus. This fungus essentially eliminated American chestnut (*Castanea dentata*) from the forest canopy in the Appalachians and Ozark chinquapin from the forests of the Ozarks. The purpose of this chapter is to describe various other threats—past, present, and potential—to the forests of the Appalachians and Ozarks.

Keywords American chestnut · Chestnut blight · Introduced plants · Introduced fungi · Overharvesting · Diseases · Ginseng · Morels

13.1 Introduction

Human activities, including both inadvertent and intentional introduction of non-native plants, animals, insects, and fungi, have had a considerable effect on forests throughout the eastern deciduous forest biome, including both the Appalachians and the Ozarks. Moreover, some of the indigenous organisms associated with these forests have been overharvested by humans. The purpose of this chapter is to discuss a number of instances (including some of them less publicized and thus less known) in which forests in the two regions considered in this book have been and will continue to be affected negatively by activities or circumstances for which humans were directly or indirectly responsible. However, only a few of the many past and ongoing threats to the forests are highlighted in this chapter. Novel or recurring threats are constantly emerging and remain somewhat unpredictable. Moreover, it is important to realize that most threats to forests are a combination of both biotic and abiotic factors.

S. L. Stephenson, R.-M. Muzika, *Forests of the Southern Appalachian Highlands and Interior Highlands of the United States*, Plant and Vegetation 20,
https://doi.org/10.1007/978-3-032-01155-8_13

13.2 Diseases

13.2.1 Chestnut Blight

In 1904, when Herman W. Merkel, a forester at the New York Zoological Park in Bronx Park, noticed dying chestnut trees, he could hardly have known that this event would portend what has been described as the worst ecological disaster in forest history (Merkel 1906; Stephenson et al. 1993). The bark on the dying trees had developed obvious cankers (necrotic depressions), which indicated the presence of some kind of disease. William A. Murrill of the New York Botanical Garden identified the disease organism as a fungus, which he called *Diaporthe parasitica.* Taxonomic considerations ultimately led mycologists to rename it, first as *Endothia parasitica* and then as *Cryphonectria parasitica.* The latter designation is how the fungus is currently known.

The fungus affecting the trees had been introduced from Asia about 1900, probably on nursery stock of an Asian chestnut such as Chinese chestnut (*C. mollissima*). Asian chestnuts can be infected with the fungus but usually do not succumb to it. However, it was soon discovered that the fungus was highly virulent when infecting American chestnut. Once established in a tree, the fungus effectively girdles the stem, thus killing the tree. Within an amazingly short period of time, the fungus (referred to as chestnut bright or sometimes the chestnut bark disease) spread throughout the entire natural range of American chestnut. By about 1950, there were few chestnut trees in all of North America that had not been killed by chestnut blight. Where chestnut made up a large percentage of the forest canopy, the standing dead trees presented a truly depressing sight (Fig. 13.1). The last surviving chestnut-dominated forests probably occurred on the upper slopes of mountains in the Great Smoky Mountains National Park, where still living but diseased trees were reported in the late 1940s (Whittaker 1956). Chestnut wood is highly resistant to decay, and dead but still standing chestnut trees (sometimes referred to as “gray ghosts”) were not uncommon in the forests of the Appalachians in the 1960s and 1970s. Amazing, a very few individuals were relatively intact as late as 2023 (Fig. 13.2).

The natural range of American chestnut did not extend to the Ozarks, so the forests of the region did not experience chestnut blight in the same way as those of the Appalachians. However, although it is not widely appreciated, the chestnut blight did have some impact on the Ozarks. This is because it also affected chinquapin, a member of the same genus as American chestnut. The fungus appears to have arrived in the Ozarks later than in the Appalachians, but populations of chinquapin were greatly reduced by the early 2000s. There are two indigenous species of chinquapin found within the Appalachians and/or Ozarks. The Allegheny chinquapin (*Castanea pumila*) is found in both regions, although it is more widespread in the Appalachians. The other species, the Ozark chinquapin (*C. ozarkensis*), occurs only in the Ozarks. Allegheny chinquapin is usually a shrub or small tree. However, prior to the arrival of the blight, Ozark chinquapin sometimes could reach a diameter of

Fig. 13.1 Dead American chestnut trees in a forest in the Ridge and Valley of southwest Virginia in the early 1930s. (Photographer unknown)

Fig. 13.2 Relatively intact trunk of a dead American chestnut tree in the Great Smoky Mountains. (Photo by Fred Paillet)

60 cm and a height of 18–20 m, making it part of the forest canopy. Both species of chinquapin are associated with dry open and often rocky sites, with the largest individuals occurring on dry ridgetops. In the past, chinquapins often made up part of the vegetation of fencerows and also were not uncommon along roadsides, where their nuts (which are much smaller than those of American chestnut) were collected and eaten. Stories are told of boys taking a pocket full of chinquapin nuts with them to school and eating them during the day. Unfortunately, due to the chestnut blight disease, it would be difficult to collect the nuts in such quantities today.

Prior to the arrival of the blight, American chestnut was once a very common and economically important tree. Cochran and Braasch (1990) estimated that in pre-blight Appalachian forests, chestnut accounted for about 25% of the total number of trees present. This would mean that there would have been three to four billion trees

Fig. 13.3 Large American chestnut trees in western North Carolina in the early 1900s. (Photo from the American Lumberman, 1910)

in over the entire range of the species. Some of these trees reached a considerable size (Fig. 13.3). Wood from the chestnut tree was valued for furniture, crossties for railroads, flooring, telegraph poles, shingles, fence posts, and sometimes entire buildings, particularly barns. Chestnut was both extremely durable and easy to work, even using the relatively primitive tools available in the 1700s, 1800s, and early 1900s. Moreover, the tree was a primary source of tannin for the leather industry (Bloomfield 1983). The nuts were an important food crop for humans and wildlife alike (West 1988).

Today the American chestnut survives mainly as a minor understory shrub sprouting from the root systems of long dead trees. Persistent efforts to breed more resistant varieties of the tree have been underway for several decades but have met with limited success. At present, it seems unlikely that American chestnut will ever regain its former position in the canopy of the forests of the Appalachians.

13.2.2 Beech Bark Disease

American beech (*Fagus grandifolia*) is consistently present in moist broadleaf forests throughout the Appalachians and along major rivers in Arkansas Ozarks and in the Ouachita Mountains but not in the Missouri Ozarks. The species also occurs along with red spruce (*Picea rubens*) in some high-elevation mixed broadleaf-coniferous forests. In 1890, what was to become known as beech bark disease was first detected in eastern Canada but subsequently spread south to become established in the Appalachians. The disease originated in southern Europe, where it is now a problem for European beech. Beech bark disease is what is known as a disease complex, which means that there is more than a single organism involved. For beech bark disease, the organisms involved include a scale insect (*Cryptococcus fagisuga*) and a fungus (either *Neonectria ditissima* or *N. gaginata*). The disease occurs when the scale insect feeds on beech trees and creates fissures in the bark, allowing the fungus to enter the tree. The fungi then kill the inner bark tissue, forming cankers which can coalesce and effectively girdle the tree, leading to its death. Beech bark disease has been devastating and is exceedingly widespread. The mortality it has caused has changed the composition of those forest communities in the Appalachians in which beech is one component.

Beech can reproduce vegetatively by producing stump sprouts and root suckers. When the main stem is killed by beech bark disease, it produces secondary stems that can be numerous enough in a beech forest to create conditions (e.g., shade) unfavorable for the establishment and growth of the other species of trees present. Consequently, beech bark disease indirectly affects more than just beech.

13.2.3 Beech Leaf Disease

Beech leaf disease was first noted in 2012 on American beech trees in northeast Ohio in the United States and has since spread throughout the Northeast and into Virginia as of 2021. The causal agent is a nematode (*Ditylenchus crenatae* ssp. *mccanna*) which is possibly native to East Asia. There may be bacterial infections associated with disease as well. It spreads rapidly, but there is a lag in symptom development. Therefore, identifying its actual geographic spread can be a challenge. The presence of the disease stresses and can cause mortality in both young and old beech trees. Studies have suggested that this disease could eventually extend over the entire range of American beech. The nematode damages buds and the characteristic symptoms are dark bands between the leaf veins, browning of foliage, and mortality of branches and eventually the entire tree. Beech leaf disease represents an example of an unexpected and emerging novel disease, possibly a disease complex. As a long-lived, late successional tree species in North America, American beech represents a foundation species of mesic forest ecosystems. Ecological consequences of the loss of this species could be dramatic.

13.2.4 Dogwood Anthracnose

Dogwood anthracnose, caused by the fungus *Discula destructiva*, was first reported as a disease of flowering dogwood (*Cornus florida*) in the northeastern United States in 1978 (Holzmueller et al. 2006). The fungus involved had been inadvertently introduced from Asia and soon spread southward, reaching the Appalachians in 1987. *Discula destructiva* causes serious losses in dogwood, including all varieties of *Cornus florida*. The infection caused by the fungus begins in the leaves and spreads to the branches, which soon die. In the trunk of the tree, infected areas develop into cankers, which kill the entire tree. In some localities, dogwood mortality has reached 90%.

In the Appalachians, infection by the fungus is most likely at intermediate elevations and on moist sites. In the Ozarks, infection can occur anywhere flowering dogwood is found but also is most prevalent on moist sites. Flowering dogwood typically occurs as an understory species, and shade increases both the risk of infection and ultimate mortality. Flowering dogwood is of considerable direct ecological important to the animals of the forests in which it occurs. Numerous birds and small animals feed upon the fruits, and dogwood twigs are important source of browse for the white-tailed deer. The conspicuous floral display of flowering dogwood also marks the onset of spring in the Appalachians and Ozarks. The most immediate consequence of dogwood anthracnose is that the composition of the understory of forests where dogwood was present have shifted towards other species.

13.2.5 Dutch Elm Disease

Dutch elm disease, which gets its name from where it was first formally described and not from where it originated, is caused by microscopic fungi in the Ascomycete genus *Ophiostoma*, particularly *O. ulmi*. Dutch elm disease was first reported from the United States in 1930, having been introduced from Europe. Like chestnut blight, Dutch elm disease is a wilt disease and affects the vascular system of the infected tree. The disease is thought to have been introduced to Europe from Asia in the mid-1940s. Once introduced into the United States, Dutch elm disease proved to be devastating for species of elm, especially American elm (*Ulmus rubra*). This tree was virtually eliminated from portions of its former range. Dutch elm disease is dispersed by bark beetles. Female beetles locate dead or dying elms and excavate a space (or gallery) between the bark and wood of the trunk or branches where the beetle lays its eggs. If the spores of *Ophiostoma* are introduced to the tree by the female beetle, they germinate to produce a mycelium that infects the wood, causing the tissues to turn brown or black. The mycelium in turn produces tremendous numbers of spores. The beetle eggs hatch and ultimately develop into adult beetles. When these new adult beetles emerge through the bark to disperse, many of them carry the spores on and in their bodies, which can infect another elm tree, thus continuing the disease cycle.

13.3 Introduced Insects

13.3.1 Balsam Woolly Adelgid

The balsam woolly adelgid (*Adelges piceae*) was introduced into North America from Europe in 1908 and was first recorded from the Appalachians in 1957. This tiny, soft-bodied insect has proved to be a major threat to both balsam fir and Fraser fir throughout their respective ranges in the Appalachians and elsewhere. It is responsible for a severe decline of Fraser fir in the red spruce-Fraser fir forests of the southern Blue Ridge (Amman and Speers 1965; Busing and Clebsch 1988). When present on the leaves of fir, the insect has the appearance of a white, woolly spot, hence the name. The insect stresses the tree by feeding on phloem. After being initially infected, Fraser fir trees usually die within several years. For forests containing a high percentage of Fraser fir, this leaves behind a rather stark landscape (Fig. 13.4). Fortunately, the populations of Fraser fir in these forests to seem to possess the capability to recover to some extent. In addition to its impact on naturally occurring fir, the balsam woolly adelgid has had a major negative effect on Christmas tree farms in the Appalachians. Disjunct native populations of fir also have been affected.

13.3.2 Hemlock Woolly Adelgid

The hemlock woolly adelgid (*Adelges tsugae*) is an aphid-like insect similar to the balsam wooly adelgid and native to Asia. It was first reported from the eastern United States in the 1950s and has expanded its range to the entire Appalachians. The hemlock woolly adelgid is a serious pest for hemlocks in the Appalachians, where both eastern (or Canadian) hemlock (*Tsuga canadensis*) and Carolina hemlock (*T. caroliniana*) occur. It has extended over the entire range of Carolina

Fig. 13.4 Dead Fraser fir (*Abies fraseri*) trees in a high-elevation forest of the Great Smoky Mountains National Park in July of 2009

hemlock and about half the range of eastern hemlock. Hemlock does not occur naturally in the Ozarks. The hemlock woolly adelgid has caused tremendous damage for decades to forests from New England to the Great Smoky Mountain National Park. Once infested, a tree is likely to die in 4 to 10 years, but at times with severe infestations, mortality occurs more quickly. Among the ecological consequences of a loss of hemlock is the dramatic shifts in microclimatic along rivers and streams and restructuring of communities of organisms ranging from species of plants to species of birds. Like beech, hemlock is a very understory tolerant, late succession, and long-lived species. The loss of this foundational species has substantial, ecological, economic, aesthetic, and cultural consequences (Ellison et al. 2018).

13.3.3 Spongy Moth

The spongy moth (*Lymantria dispar*), formerly known as the gypsy moth, was introduced into Massachusetts from Europe in 1869, and the first outbreak occurred in 1889 (Jankovic and Petrovskii 2013; Rindos and Liebhold 2023). Over time, the moth spread southward and westward, reaching the northern portion of West Virginia in 1972 and the northern portion of Virginia in 1984. Only this northernmost portion of the Appalachian region has been impacted by the spongy moth thus far. Owing to a well-funded, long-term program of integrated management to slow the spread of this species, the spongy moth occurs over only about one-third of its potential range in the United States. There was an outbreak of this insect in northwest Arkansas and southwest Missouri in the 1990s, but that population was successfully eradicated. The Ozark forests are otherwise very susceptible, with the dominant tree species present among those preferred by the insect. Spongy moth larvae are known to be capable of feeding upon the foliage of at least 300 different species of plants, but they show a decided preference for certain species of trees, which often suffer extensive defoliation as a result. In extreme cases, complete defoliation takes place. Among the preferred tree species are various oaks, including chestnut oak, black oak, red oak, and white oak (Gansner and Herrick 1985).

The stress of defoliation has on a particular tree depends upon a number of factors. The most important are (1) the species of tree involved, (2) the health of the tree, (3) the extent of the defoliation, and (4) moisture conditions of the site where the tree occurs. Various chemical insecticides including lead arsenate, DDT, orthene, carbaryl, and the insect growth regulator Dimilin® along with biological insecticides such as *Bacillus thuringiensis* (Bt) and Gypchek have been used in efforts to suppress and eradicate the spongy moth. These have had varying degrees of success. The use of DDT for this species and also the brown moth, a non-native species that was once an important defoliator but has since nearly vanished, was the topic of a chapter in Rachel Carson's classic book, *Silent Spring*.

13.3.4 Emerald Ash Borer

The emerald ash borer (*Agrilus planipennis*) is a phloem-feeding beetle in the metallic wood boring beetle family and was inadvertently introduced into the United States from eastern Asia in the early to mid-1990s, with established populations discovered in 2002 (Flower et al. 2013). This beetle relies solely on ash trees (*Fraxinus* spp.) to complete its life cycle. The most common species of ash in the forests of the Appalachians and Ozarks is white ash (*F. americana*), which often makes up a significant component of forest communities. Other species of ash found in one or both regions are pumpkin ash (*F. profunda*), black ash (*F. nigra*), green ash (*F. pennsylvanica*), blue ash (*F. quadrangulata*), and Carolina ash (*F. carolina*). However, none of these is nearly as widespread and common as white ash. In 2022 alone, the emerald ash borer affected 581,963 ha (over 1.4 million acres) of forest land (Pandat and Conkling 2024).

The feeding activities of the larval stage of the emerald ash borer effectively girdle the tree by destroying the phloem tissue, resulting in the mortality of an individual ash tree in as little as 2 years and mortality of a local population in as little as 5 years. Unfortunately, the emerald ash borer has the potential to be dispersed over relatively long distances, both through natural dispersal and inadvertently through the transport of firewood, allowing it to spread rapidly over the natural range of *Fraxinus*. The loss of *Fraxinus* from the canopy will have the same effect as the loss of American chestnut. Non-ash tree species will expand into the habitat spaces in the canopy that ash once occupied, and species composition of many forests will be altered. Regeneration of ash occurs in forests where overstory ash trees have been decimated, but analysis suggests that few ash trees survive to regenerate (Ward et al. 2021).

13.4 Hogs

European hogs (*Sus scrofa*) were probably introduced into the Ozarks in 1541, when the Spaniard Hernando de Soto led an expedition that is thought to have passed through part of the Arkansas River Valley. The expedition had a herd of hogs that represented a reliable source of food. Female hogs had litters, which allowed the number of hogs to be maintained. However, some hogs undoubtedly escaped to reestablish free-ranging populations. Domestic hogs were introduced to the Appalachians in the early 1700s by settlers. In many instances, the settlers allowed the hogs to range free in the forests, where they could feed upon acorns and various plants. Some of these hogs soon established feral populations in the region.

The European wild boar was intentionally introduced into both the Appalachians and Ozarks in order to provide big-game hunting opportunities. Since a large male hog can weigh as much as 270 kg, it certainly qualifies as a big game. European wild boars were introduced into eastern Tennessee in 1962 and southwestern

Virginia in 1971. They were already present in the Ozarks well before this. Hybrids between the European wild boar and domestic hogs are now common, so a feral hog encountered in the wild is likely to be a hybrid.

Wild boars were inadvertently introduced into the Great Smoky Mountains National Park about 1920, and a population has existed in the park since then. Although wild boars provide opportunities for hunting, they also disrupt the habitats in which they occur. Soon after wild boars were present in the Great Smoky Mountains, it was noticed that their rooting was destructive to the grassy balds that occur in the park. The same was true for forests throughout the park. Moreover, they fed upon plants and small animals, depressing populations of the species in question. Consequently, feral hogs—either the European wild boar, the domestic hog or a hybrid of the two—represent a significant threat to the forests in which they occur.

Feral hogs represent a substantial concern in the Ozarks where population are highest in southern Missouri. They are also common in southern Arkansas, although that does not fall within the Ozarks. Feral hogs cause considerable disturbance to natural areas and disrupt species composition of plant communities, particularly sensitive glade habitats. They transmit diseases and are prolific breeders. In the Appalachians, hogs are most common in the Southern Blue Ridge. Hunting and eradication programs have helped reduce numbers, but hogs still remain a threat to forests in both the Appalachians and Ozarks.

13.5 Introduced Plants

13.5.1 Garlic Mustard

Garlic mustard (*Alliaria petiolata*), a biennial herbaceous plant native to Europe, Asia, and parts of northern Africa, was introduced to the United States in the mid-1800s by settlers from Europe. Unlike most of the other organisms discussed in this chapter, this plant was intentionally introduced, since it was used, among other things, as a source of a garlic-like flavoring in foods and medicinally as a diuretic. Although at first limited to gardens, garlic mustard quickly spread to other areas, ultimately becoming naturalized throughout most of the eastern United States.

As a biennial, garlic mustard requires 2 years to complete its life cycle. In the first year, the plant exists as a rosette of leaves close to the ground. In the second year, the mature flowering plant is produced. The seeds of garlic mustard are very small and easily dispersed. The plant is unpalatable to most native herbivores and is capable of surviving in the forest understory, where its presence results in a reduction in overall diversity of the plants present. The competitive success of garlic mustard in such situations appears to be due to its production of allelochemical substances that suppress mycorrhizal associations in neighboring non-garlic mustard plants (Cipollini 2016).

Since the garlic mustard seeds are so small, they are easily dispersed, sometimes for long distances by a suitable vector. The plant often tends to invade a new area

Fig. 13.5 Garlic mustard (*Alliaria petiolata*) growing along a roadside in the Ridge and Valley of southwest Virginia. (Photo by Steven L. Stephenson)

along the ecological corridors represented by roads and streams, advancing each year. An example of garlic mustard plants growing along a highway in the Ridge and Valley of southern Virginia is shown in Fig. 13.5.

13.5.2 Multiflora Rose

Multiflora rose (*Rosa multiflora*) is a shrub native to eastern Asia. In 1866, it was intentionally brought to the United States as a rootstock for ornamental roses. The plant grows well on poor sites, and it was later used for erosion control and as a source of food for wildlife, since it produces fruits that are readily consumed by birds. Unfortunately, when the birds consume the fruits, they disperse the seeds in their droppings, which inadvertently spreads multiflora rose to new sites.

As it grows, multiflora rose forms dense almost impregnable thickets that degrade the overall habitat in which they occur by outcompeting native species and reducing regeneration of tree species. Further complicating this issue, the leaves of multiflora rose are not readily consumed by native herbivores, and there are no natural controls that would limit the spread of the plant. Although more commonly associated with roadsides and abandoned fields, multiflora rose also can invade the understory of forests.

13.5.3 Kudzu

Kudzu (*Pueraria montana*) is perennial vine native to Japan and southeast China. It was first introduced to the United States during the Philadelphia Centennial Exposition in 1876, where it was considered to be an ornamental plant with

Fig. 13.6 Dense growth of kudzu (*Pueraria montana*) along a highway in the Ridge and Valley of southwest Virginia. (Photo by Steven L. Stephenson)

sweet-smelling flowers. From the 1930s through the 1950s, the Soil Conservation Service of the United States promoted kudzu for its use as an effective control for soil erosion, and it was planted extensively throughout the southeastern United States. However, it was discovered kudzu could grow so rapidly and abundantly that it could overcome anything in its path. The plant spreads through long stems that root at the tip when they come in contact with the ground and by aerial stems that root at the nodes to form new plants. Once established, kudzu grows at a rate of about 30 cm/day, and mature plants can extend more than 30 m.

Kudzu is typically found along roadsides, where it can climb up any trees that are present (Fig. 13.6). However, the plant sometimes can extend well away from a road and literally overrun a forest community, especially one that is rather open and with relatively few trees. Kudzu dies back in the winter, leaving behind a dense brown mass of dead stems and leaves. A thick growth of kudzu is essentially a monodominant community—few plants can grow beneath it. However, kudzu is a member of the family Fabaceae and thus is a nitrogen-fixing plant. As such, it actually enriches the soil in which it grows. Kudzu is characteristically found only at lower elevations in the Appalachians but its distribution is not limited by elevation in the Ozarks.

13.5.4 Japanese Honeysuckle

Many non-native plants have become so naturalized in the United States that most people don't recognize the fact that were not part of the original flora. One of the best examples is Japanese honeysuckle (*Lonicera japonica*), a climbing vine introduced into the United States in the early 1800s as an ornamental plant because of its beautiful and fragrant flowers. However, Japanese honeysuckle is classified as an aggressive invasive capable of altering habitats. Although most common along roadsides, fencerows, and forest edges, this plant can invade and ultimately overgrow forests. Because Japanese honeysuckle is only moderately shade tolerant, only forests disturbed by logging or some other disturbance are likely to be susceptible to Japanese honeysuckle.

Japanese honeysuckle is only one of several invasive members of the genus *Lonicera* found in the Appalachians and Ozarks. Bush honeysuckle (*L. maackii*) and tatarian honeysuckle (*L. tatarica*) are shrubs, not vines like Japanese honeysuckle. Each can grow several meters high and out compete native plants. These other honeysuckles put out their leaves earlier than most native plants in the same habitat, which provides shade that is detrimental to neighboring plants. Buch honeysuckle is capable of surviving under full shade, which allows it to invade forests. Otherwise, both bush honeysuckle and tartarian honeysuckle tend to be most common along roadsides.

13.5.5 Tree of Heaven

Tree of heaven (*Ailanthus altissima*), also called paradise tree, is a medium-sized tree native to temperate areas of China. It was brought to the United States in 1784 for use as an ornamental. The tree was once a favored ornamental and often planted along streets during the 1800s and early 1900s (Sladonja and Sušek 2015), in part because its tolerance of pollution. It still retains some value as an ornamental, but when tree of heaven escaped cultivation and became established in natural habitats, it soon became recognized as one of the most problematic invasive woody plants in the United States.

The tree can be easily recognized by its large pinnately compound leaves, smooth gray bark, and distinct odor (the Chinese name is "stinking tree"). It was used in traditional Chinese medicine in China, but the primary use in the United States is as a poor type of firewood. Tree of heaven grows very rapidly, and the tree produces allelochemicals which inhibit the growth of other plants. The ability of tree of heaven to regrow from roots makes it exceedingly difficult to eradicate when it forms dense clones.

13.5.6 Mimosa

Mimosa (*Albizia julibrissin*), also called silk tree, is a small to medium-sized tree that was introduced to South Carolina in 1785 as an ornamental. When maintained in an urban setting such as in a lawn or park, the bipinnate frond-like leaves and pink feathery flowers of mimosa make it very attractive option, and mimosa is still planted despite the fact that in nonurban settings it is an undesirable invasive. The plant produces numerous seeds, and these are readily consumed by birds. This increases the chances of it being dispersed to new localities. In nature, mimosa is most likely to be found along roadsides and near streams. The tree grows very rapidly and can overtop surrounding plants. However, mimosa trees are short-lived and usually do not reach an age of 30 years.

13.5.7 Princess Tree

The princess tree (*Paulownia tomentosa*) was introduced into the United States from China in the 1840s as use as an ornamental tree, something that is still the case. In the spring, the princess tree produces clusters of large fragrant violet-blue flowers. This makes the tree very attractive to landscapers and home owners. However, the tree also produces small, readily dispersed seeds, is exceedingly fast growing, and can survive on poor sites. In fact, the princess tree has received some use in reclamation of surface mines. The species has successfully escaped cultivation and now has become naturalized throughout the eastern United States. It is most common on disturbed sites and frequently occurs along forested roadsides, where it is easy to spot when the tree is in flower. Princess tree is found throughout the Appalachians and Ozarks, although it is limited to lower elevations in the Appalachians.

13.5.8 Japanese Stiltgrass

The introduction of nonindigenous species to a new continent can occur in a number of ways, but Japanese stiltgrass (*Microstegium vimineum*) apparently came to the United States in a most unusual way—as packing material for fragile dishes shipped from China (Judge et al. 2005). In Asia, the plant was commonly used as packing material and has sometimes been referred to as "packing" grass. Japanese stiltgrass is low-growing annual grass that spreads over the ground as it grows. The plant produces numerous tiny seeds that are readily dispersed. It typically occurs in moist and shady habitats. Over those portions of the Appalachians and Ozarks where Japanese stiltgrass occurs, it is frequently found along woodland trails and roads. Unlike most grasses, this invasive plant is capable of surviving beneath the forest canopy, where it out compete native plants.

13.5.9 Crown-Vetch

Crown-vetch (*Secuorigea varia* but sometimes listed as *Coronilla varia*) is a low-growing vine native to Africa, Asia, and some parts of Europe. The plant was introduced in the United States in the 1800s for the purpose of providing control of erosion and as ground cover. Crown-vetch produces clusters of attractive rose-pink flowers in late spring and early summer, and these are a familiar sight along interstate highways and other roads. Although once promoted for its use as a ground cover for roadsides, this plant is capable of such rapid growth that it is difficult to control once established.

Crown-vetch, like kudzu, is a member of the Fabaceae, which means that it is a nitrogen-fixing plant. This is a positive factor for the habitats in which it occurs since surrounding soils are enriched. However, areas on which crown-vetch has become established tend to be monodominant, with few other plants present.

13.6 Disease Complexes

13.6.1 Oak Decline

Most tree diseases can be attributed to a single causal agent, although there are exceptions that involve the combined effects of two different agents (e.g., beech bark disease). Oak decline, which was first noticed in the late 1800s and has become more prevalent since the 1950s, is more complicated. During 2022 oak decline was identified as the mortality agent accounting for 25,391 ha (62,742 acres) of mortality in the eastern United States. The most obvious symptoms of oak decline are a reduction in growth rates and dieback in the branches of the affected tree, but oak regeneration and mast (acorn) production are also affected (Biocca et al. 1993). Over time, dead branches become conspicuous. Oak decline has been the focus of considerable research, but the exact causes are still disputed. It is likely that many factors can be involved, and what we recognize as oak decline involves a number of overlapping factors. Among the most important of these is related to site conditions since oak decline tends to be more prevalent on lower productivity sites. Oaks that experience oak decline commonly occur at upper slope positions, which tend to be rocky and relatively xeric. Weather also is likely to play a role since drought conditions in a given year (or years) place additional moisture stress on trees. Moreover, species of oak differ with regard to their "susceptibility" to oak decline. Members of the white oak group are less likely to experience oak decline than members of the red oak group. In the Ozarks, oak decline often occurs in sites that were formerly dominated by shortleaf pine prior to the extensive logging of that species. Age of the tree and the community in which it occurs are important as well. A number of secondary factors can be involved, including such things as wood boring beetles and Armillaria root disease. Another factor is more difficult to evaluate. Oak trees are ectomycorrhizal and thus dependent upon the fungi associated with their root system. Anything affecting the fungi involved would have the potential of affecting the host tree. Also, trees, like all other plants, are subject to being infected by various fungi, whose presence is difficult to detect unless the fungus produces fruiting bodies on the outside of the tree (if indeed it produces macroscopic fruiting bodies). Some fungi can be present in a tree for years without revealing their presence. Finally, oak decline is almost invariably associated with older trees, which presumably would be more susceptible to any type of stress.

13.6.2 Spruce Decline

In the early 1980s, Siccama et al. (1982) reported that dendroecological (tree-ring) studies of red spruce in the mountains of Vermont indicated that this species had exhibited reduced growth over the past two decades. It was soon suggested that this decline could be attributed to a gradual rise in the levels of acid deposition from industrial sources (Johnson and Siccama 1983). However, some scientists suggested that it was a response to other factors, including drought, age, insects, and fungal pathogens. Later, Adams et al. (1985) and others reported the same pattern for reduced growth in the mountains of the southern Appalachians. The decline of red spruce in this region appears to have been initiated by repeated winter damage to needles and buds from freezing or desiccation. The red spruce trees demonstrated the same pattern of narrow growth rings observed in the northern Appalachians.

The decline of red spruce is one of the best documented cases of the effect of acid deposition on a sensitive species. In fact, the data on red spruce decline were used as part of the scientific justification behind reductions in pollution mandated by the 1990 Amendments to the Clean Air Act. However, after decades of reductions in acid deposition, some more recent reports (Kosiba et al. 2017) have indicated that many red spruce trees have exhibited increased growth since 2001, which suggest that red spruce is rebounding from historical acid deposition stress.

Most of the populations of red spruce in the Appalachians occur in the southern portion of the region, more specifically in the southern Blue Ridge. Well-established populations are found at a limited number of places in the northern portion. These are located in the northern Blue Ridge, the northern Ridge and Valley, and on the Appalachian Plateau, where they are likely to be near the lower elevational limit for red spruce. During the Holocene Climatic Optimum, due to the higher temperatures, red spruce would not have occurred some of these places. The increase in temperatures projected for global climate change is expected to shift the distribution of red spruce to higher elevations, which don't always exist in the northern portion of the Appalachians.

Concerns about spruce mortality in the Appalachians, specifically West Virginia. arose as early as the 1880's (Hopkins 1899). Large numbers of red spruce trees in a portion of the Appalachians appeared to have died in the late 1800s. In 1891, Charles Millspaugh, a botanist with the West Virginia Agricultural Experiment Station, published some observations he had made on a visit to Cheat Mountain in 1890.

> Thousands and thousands of acres of what much have once been a magnificent black-green forest of living spruce, was now viewed as an immense waste of dead and dying trees, presenting a desolate and dreary landscape.

> From my own observations, I would judge that forty percent of the trees were dead of the 15,000 to 20,000 acres near Cheat Bridge. The general supposition by people living here is that the timber died from the effect of the drought of 1882, some claimed that the trees had come to maturity and were dying from old age, giving their reasons that the larger trees seemed to die worse than the smaller ones, while a very few thought that it might have been caused by insects.

Since Millspaugh made no mention of any evidence of previous fires, the assumption is that something else must have killed the red spruce trees. This region of the Appalachians was soon to be logged, so whatever factor or agent was responsible for the death of red spruce will never be known. However, Hopkins (1899) identified many factors such as drought stress, fires, and spruce beetles as well as other bark beetles and borers as possible causes of mortality,

13.6.3 Armillaria Root Disease

Armillaria root rot is a fungal disease caused by several different species in the genus *Armillaria*, although *A. mellea* is the species most commonly cited as a causal agent in tree mortality. Although sometimes called "oak root fungus," *Armillaria* has no specificity for oaks and can infect numerous other plants. When Armillaria root rot infects a tree, it kills the vascular cambium, the tissue responsible for producing new bark and wood, in the roots of the tree. It then spreads laterally to the main stem and in time can girdle the base of the tree and kill it. As such, the fungus would be classified as a pathogen. However, *Armillaria* is also a wood-decay fungus and breaks down the woody tissues of the tree. When the host tree dies, the fungus can survive for an extended period of time in the dead wood of the trunk but remaining capable of infecting another tree. This dual nature of *Armillaria* makes it difficult to manage. Symptoms of Armillaria root rot are variable, ranging from stunted leaves to chlorotic needles and dieback of twigs and branches. However, it often produces conspicuous fruiting bodies (mushrooms) at or near an infected tree. These occur in tight clusters and are light brown (the color of honey, which account for the common name "honey mushrooms"). Another identifying feature of an *Armillaria* infection is the presence of black string-like structures (called rhizomorphs) found beneath the bark of infected trees or decaying logs in which the fungus is present. *Armillaria* is a gilled mushroom like the commercial button mushroom (the one frequently found on salad bars). Although considered edible, *Armillaria* should not be collected for the table because there are several poisonous look-alike mushrooms that are found in similar situations.

Armillaria mellea is unlike most of the diseases covered in his chapter. It is an indigenous fungus and was not introduced from outside North America. However, the fungus is largely restricted to the eastern United States (as well as the Lake States and the Midwest) and is commonly encountered in both the Appalachians and Ozarks. Viewed objectively, *Armillaria mellea* is an ecologically important component of forest ecosystems, playing a critical role in wood decomposition and nutrient recycling. *Armillaria* is typically implicated as a mortality agent in oak decline.

13.6.4 *Phytophthora* Root Disease

It is difficult to characterize both the organism *Phytophthora* as well as the manifold diseases it causes. Although often confused as a fungus, *Phytophthora* (Greek for plant destroyer) is a water mold (phylum Oomycota), a fungus-like organism that is known to cause a type of root disease. Possibly the most relevant water mold in human history was *P. infestans*, the cause of the potato blight and resultant great Irish famine. The globally distributed *Phytophthora cinnamomi* has been the most studied species in this genus, followed by *P. ramorum*, which is the causal agent of sudden oak death. *Phytophthora cinnamomi* has been known to cause massive tree death in Australia but also has been identified in North America, where it was associated with historic losses of conifers in the southeastern United States and more recently with mortality of oaks in portions of the Appalachians and Ozarks (Reed et al. 2019). It is likely not native to North America despite its longevity on the continent. Moreover, there is an ever-increasing number of new species of *Phytophthora* being identified (Jung et al. 2024). Before the damage by chestnut blight in the early twentieth century, *P. cinnamomi* was identified as a major source of mortality of American chestnut trees, particularly in the southern Appalachians, and remains a concern for restoration of chestnut (Gustafson et al. 2022).

Phytophthora is spread through the soil, and therefore the "spores" can be carried on hiking boots. Consequently, proximity to walking trails has been identified a risk factor for introduction of this organism to a particular site. Although *P. cinnamomi* tends to be restricted to regions south of 40 °N latitude, there is concern that increasing global temperatures will cause a northward shift and render more forests susceptible to *Phytophthora* infection. Generally speaking, *Phytophthora* is among the most damaging plant pathogens across the globe. Although it is widespread globally, the continued spread of this pathogen in North America is likely due to climate change, and this will exacerbate forest health conditions.

13.7 Overharvesting

13.7.1 Bryophytes

As pointed out by Studlar and Peck (2007), industries that involve nontimber forest products are notoriously under-reported and under-investigated (Lynch et al. 2004). One notable example is the commercial moss harvest trade. It has become clear that harvesting green mosses (as opposed to *Sphagnum* or peat mosses) is a multi-million dollar industry in the United States with two regional centers. The first of these is the Pacific Northwest and the second is the Appalachians (Muir et al. 2006). Most studies have focused on the Pacific Northwest, where the commercial moss harvest trade is reasonably well documented. This is not the case for the Appalachians,

Fig. 13.7 Boulders covered with mats of mosses in a high-elevation oak forest in the Ridge and Valley province of southwest Virginia. (Photo by Steven L. Stephenson)

where the total quantity of moss harvested is unknown because harvest records for private lands (such as timberlands and mined lands) are not available. Forest Service records and anecdotal reports indicate a substantial historic unregulated moss harvest on the Monongahela National Forest in West Virginia (Studlar and Peck 2007), where the mosses harvested are used rather extensively as a decorative material in the floral trade.

Bryophytes (mosses and also liverworts) are often overlooked in forest ecosystems, although this is difficult to do in the high-elevation forests of the Appalachian, where the cover of bryophytes can often form an almost complete cover over the forest floor and large boulders (Fig. 13.7). These bryophytes play significant roles in ecosystem processes (e.g., nutrient and water cycling) while also providing habitats for numerous invertebrates as well as some small vertebrates such as salamanders. Some fungi are associated only with mats of bryophytes, and removing a cover of bryophytes from a dead log or stump affects fungal decomposition of the wood.

13.7.2 Lycopods

Lycopods (mostly species of *Lycopodium* and related genera), commonly called club mosses, are primitive vascular plants and not mosses. On occasion, stoloniferous species (e.g., running cedar [*Diphasiastrum digitatum*]) that occur in the Appalachians have been collected and used to make a Christmas wreath. At higher elevations in the region, lycopods are abundant and their populations are unlikely to be threatened by wholesale collecting. However, a lower elevations and also for all of the Ozarks, lycopods have a more limited occurrence, and a local population could be wiped out with overzealous collecting.

13.7.3 *Orchids*

Orchids produce some of the most beautiful flowers of any plant. Orchids are particularly diverse in the Appalachians, but some colorful species also occur in the Ozarks. The unfortunate natural tendency of most people (especially children) is to pick an orchid flower when they encounter an especially colorful or intricate example. The usual fate of the flower is to be left behind on the forest floor. Most species of orchids are rare, and removing the flower doesn't allow the plant to produce seeds. This increases the likelihood of the local population not surviving. This has been documented for local populations of particular species in both the Appalachians and the Ozarks. Notable species of native orchids include the pink lady's slipper and the yellow lady's slipper.

13.7.4 *Ginseng*

Asian ginseng (*Panax ginseng*) is a perennial herb with a root that has been used for medicinal purposes in China and other Asian countries for more than 5000 years (Fig. 13.8). Some groups of Native Americans made similar use of a closely related plant (*P. quinquefolius*), known as American ginseng. During the early 1770s, colonists in southern Canada became aware of the value of this plant. Soon thereafter, American ginseng became a trade community, with ginseng roots being shipped to China and other Asian countries. By about 1750, ginseng also was being exported from the Appalachians.

Ginseng (sometimes called "sang" locally) is a characteristic plant of moist broadleaf forests in the Appalachians and the Ozarks. Ginseng roots have considerable monetary value, with a dried pound of roots from wild plants selling for several hundred (or more) dollars. Because of its high value, ginseng is subject to being over collected, and the plant has been extirpated in some localities because of

Fig. 13.8 American ginseng (*Panax quinquefolius*) with the distinctive red fruits present. (Photo by Steven L. Stephenson)

unauthorized poaching. Studies have shown that deer browse represents another concern for ginseng and that both protection from deer browse and harvesting may be necessary to maintain viable populations of ginseng in both regions (Farrington et al. 2009; McGraw et al. 2013).

The name "ginseng" is derived from a Chinese term ("jen-shen") that translates as "in the image of a man." Ginseng roots that bear some resemblance to the shape of the human body are considered highly desirable in China and command a very high price. Ginseng is sometimes confused with wild sarsaparilla (*Aralia nudicaulis*), which often grows in the same types of habitats but is much more frequently encountered. Dwarf ginseng (*P. trifolius*), a second species of *Panax* found in the Appalachians, is much smaller and is not collected in the same way as ginseng. This species does not occur in the Ozarks. Ginseng is not the only plant in the Appalachians for which there is a commercial market. Other examples include black cohosh (*Actaea racemosa*), goldenseal (*Hydrastis canadensis*), and St. John's wort (*Hypericum perforatum*).

13.7.5 Morels

As mentioned in an earlier chapter, a number of easily recognizable species of macrofungi ("mushrooms" to most people) are found in both the Appalachians and Ozarks. Some of these are heavily collected. Morels represent the best-known and most frequently harvested example (Fig. 13.9). Morels produce their fruiting bodies in the spring, well before those of other fungi appear. In many areas in both regions, numerous people visit forests during April and May to look for morels. The question has been raised by some mycologists as to the possible effect of collecting in the same locality year after year. Does it result in a drop in the number of morels? The fruiting body of a morel is the source of the spores that enable this macrofungus to reproduce. In theory, removing fruiting bodies reduces the number of spores

Fig. 13.9 Morels (including members of the genus *Morchella*) are commonly collected from the forests of the Appalachians and Ozarks during spring. (Photo by Steven L. Stephenson)

produced and presumably would lead to a long-term reduction in the population size of morels. However, essentially no data exist that would allow this situation to be evaluated.

13.7.6 Native Species

There are countless species of pathogens in a number of different groups of organisms that have caused impressive mortality in forest ecosystems in the Appalachians and Ozarks. For example, southern pine beetle (*Dendroctonus frontalis*) populations rose dramatically during the 1960s to the 1990s and then into the early twenty-first century in the southern Appalachians, but this insect essentially has been nondetectable more recently. Outbreaks of this tree-killing beetle were evident in the 1970s and 1990s in the Oauchita Mountains of Arkansas but barely noticeable since. In striking contrast, the native bark beetles of the western United States have accounted for many millions of hectares of tree mortality during the most recent decades. One type of wood boring beetle, the red oak borer (*Enaphalodes rufulus*), typically a secondary pest and not a direct tree killer, experienced an outbreak population in the Ozarks in the early twenty-first century.

In summary, both biotic and abiotic factors pose continual stress to forest ecosystems. Although it is not possible to describe all past, current or potential threats to forests, this chapter identified some of the more important and distinct concerns. Increasingly relevant is the role of climate change and human activity on forest health. Biological invasions can be driven by climate change, and the life histories of species of native insect species and their distribution are altering with changing condition and can pose increasing risk. Moreover, the rise of novel and emergent threats remains a continual concern.

References

Adams HS, Stephenson SL, Blasing TJ, Duvick DN (1985) Growth-trend declines of spruce and fir in mid-Appalachian subalpine forests. Environ Exp Bot 25:315–325

Amman GD, Speers CF (1965) Balsam woolly aphid in the Southern Appalachians and what can be done? J Forest 104:21–26

Biocca M, Tainter FH, Starkey DA, Oak SW, Williams JG (1993) The persistence of oak decline in the western North Carolina Nantahala Mountains. Castanea 58:178–184

Bloomfield (1983) What about the chestnut. Am Forests 89(4):22–23

Busing RT, Clebsch EEC (1988) Fraser fir mortality and the dynamics of a Great Smoky Mountains fir-spruce stand. Castanea 53:177–182

Cipollini D (2016) A review of garlic mustard (*Alliaria petiolata*, Brassicaceae) as an allelopathic plant. J Torrey Bot Soc 143:329–348

Cochran MF, Braasch G (1990) Chestnuts: back from the brink. Natl Geogr 177:128–140

Ellison AM, Orwig DA, Fitzpatrick MC, Preisser EL (2018) The past, present, and future of the hemlock woolly adelgid (*Adelges tsugae*) and its ecological interactions with eastern hemlock (*Tsuga canadensis*) forests. Insects 9:172

Farrington SJ, Muzika R-M, Drees D, Knight TM (2009) Interactive effects of harvest and deer herbivory on the population dynamics of American ginseng. Conserv Biol 23:719–728

Flower CE, Knight KS, Gonzalez-Meler MA (2013) Impacts of the emerald ash borer (*Agrilus planipennis* Fairmaire) induced ash (*Fraxinus* spp.) mortality on forest carbon cycling and successional dynamics in the eastern United States. Biol Invasions 15:931–944

Gansner SP, Herrick OW (1985) Host preferences of gypsy moth on a new frontier of information, USDA Forest Service, Northeastern Forest Experimental Station, Forest Research Note NE-330. USDA Forest Service, Broomall, PA

Gustafson EJ, Miranda BR, Dreaden TJ, Pinchot CC, Jacobs DF (2022) Beyond blight: *Phytophthora* root rot under climate change limits populations of reintroduced American chestnut. Ecosphere 13(2):e3917

Holzmueller E, Jose S, Jenkins M, Camp A, Long A (2006) Dogwood anthracnose in eastern hardwood forests: what is known and what can be done? J Forest 104:21–26

Hopkins AD (1899) Report on investigations to determine the cause of unhealthy conditions of the spruce and pine from 1880–1893, West Virginia Agricultural Experiment Station Bulletin 56. West Virginia Agricultural Experiment Station, Morgantown

Jankovic M, Petrovskii S (2013) Gypsy moth invasion in North America: a simulation study of the spatial pattern and the rate of spread. Ecol Complex 14:132–144

Johnson AH, Siccama TG (1983) Acid deposition and forest decline. Environ Sci Technol 17:294A–305A

Judge CA, Neal JC, Derr JF (2005) Preemergence and postemergence control of Japanese stiltgrass (*Microstegium vimineum*). Weed Technol 19:183–189

Jung T, Milenković I, Balci Y, Janoušek J, Kudláček T, Nagy ZÁ, Jung MH (2024) Worldwide forest surveys reveal forty-three new species in *Phytophthora* major clade 2 with fundamental implications for the evolution and biogeography of the genus and global plant biosecurity. Stud Mycol 107:251–388

Kosiba AM, Schaberg PG, Rayback SA, Hawley GJ (2017) Comparative growth trends of five northern hardwood and montane tree species reveal divergent growth trajectories and response to climate. Can J For Res 47:743–754

Lynch KA, Jones ET, McLain RJ (2004) Nontimber Forest product inventorying and monitoring in the United States: rationale and recommendations for a participatory approach. Report to the National Commission on science for sustainable forestry by the Institute for Culture and Ecology, Portland, OR

McGraw JB, Lubbers AE, van der Voort M, Mooney EH, Furedi MA, Souther S, Turner JB, Chandler J (2013) Ecology and conservation of ginseng (*Panax quinquefolius*) in a changing world. Ann N Y Acad Sci 1286:62–91

Merkel HW (1906) A deadly fungus on the American chestnut. NY Zool Soc Ann Rpt 10:97–103

Millspaugh CF (1891) The black spruce. West VA Agric Station Rep 3:98–101

Muir PS, Norman KN, Sikes KG (2006) Quantity and value of commercial moss harvest from forests of the Pacific Northwest and Appalachian regions of the U.S. Bryologist 109:197–214

Pandat K, Conkling BL (2024) Forest health monitoring: national status, trends, and analysis 2023, USDA Forest Service General Technical Report WO-105. USDA Forest Service, Washington, DC

Reed SE, English JT, Muzika R-M (2019) *Phytophthora* species detected in two Ozark forests with unusual patterns of white oak mortality. Plant Dis 103:102–109

Rindos M, Liebhold AM (2023) The spongy moth, *Lymantria dispar*. Curr Biol 33(12):R665–R668

Siccama TG, Bliss M, Vogelmann HW (1982) Decline of red spruce in the Green Mountains of Vermont. Bull Torrey Bot Club 109:162–168

Sladonja B, Sušek M (2015) Review on invasive tree of heaven (*Ailanthus altissima* (mill.) Swingle). Conflicting values: assessment of its ecosystem services and potential biological threat. Environ Manage 56:1009–1034. https://doi.org/10.1007/s00267-015-0546-5

Stephenson SL, Ash AN, Stauffer DF (1993) Appalachian oak forests. In: Martin WH, Boyce SC, Echternacht AC (eds) Biodiversity of the Southeastern United States: upland terrestrial communities. John Wiley and Sons, New York, pp 255–303

Studlar SM, Peck JE (2007) Commercial moss harvest in the Appalachian Mountains of West Virginia: targeted species and incidental take. Bryologist 110:752–765

Ward SF, Liebhold AM, Morin RS, Fei S (2021) Population dynamics of ash across the eastern USA following invasion by emerald ash borer. For Ecol Manag 479:118574. https://doi.org/10.1016/j.foreco.2020.118574

West RF (1988) Tolling the chestnut. Am Forests 94:76–77

Whittaker RH (1956) Vegetation of the Great Smoky Mountains. Ecol Monogr 26:1–80

Chapter 14
Climate Change

Abstract There is a mounting body of evidence that anthropogenic climate change will have a significant impact on the forests and non-forest habitats of the Appalachians and Ozarks. Even under conservative scenarios, future changes in climate are likely to include further increases in temperature with significant drying (drought) in some regions and increases in the frequency and severity of extreme weather events. Among those habitats most likely to be affected are wetlands, xeric habitats, and those associated with the highest elevations in the Appalachians. The magnitude of the changes that will take place remains unknown, but some of the expected effects are listed and discussed in this chapter.

Keywords Carbon sinks · Elevated CO_2 · Holocene climatic optimum · Keeling curve · Microbial loop · Subalpine habitats · Wetland habitats

14.1 Introduction

Climate is the primary factor determining the geographical distribution of organisms on the Earth, and this is especially true for plants (including forest trees). Particular species of plants are adapted to a range (niche) of climatic conditions, and climate change has the potential to cause shifts in the geographical distribution of these climatic niches (Parmesan 2006). In the example described in the next section, one forest type was displaced to significantly higher elevations in the Appalachians as a result of the higher mean temperatures that existed during one period of time in the past. Any change in the geographical distribution of a plant can have broad implications for other species that rely on the plant for food, shelter, or as a host (for parasitic or pathogenic organisms).

Anthropogenic climate change is likely have a significant impact on the forests and non-forest habitats of the Appalachians and Ozarks. The magnitude of the changes that will take place remains unknown, but there are enough data available to suggest the likely nature and direction of these changes. First of all, the change in

S. L. Stephenson, R.-M. Muzika, *Forests of the Southern Appalachian Highlands and Interior Highlands of the United States*, Plant and Vegetation 20,
https://doi.org/10.1007/978-3-032-01155-8_14

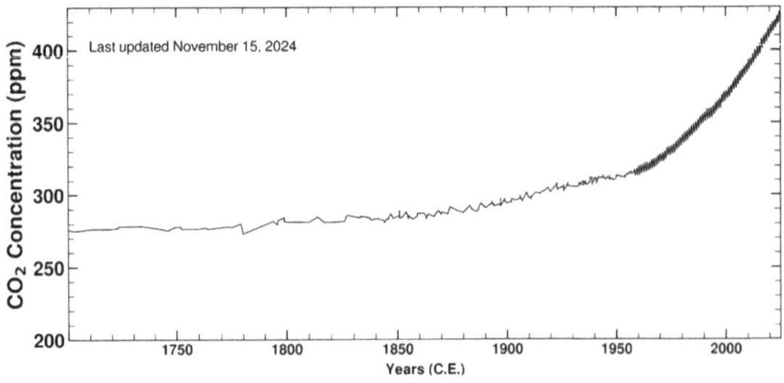

Fig. 14.1 Concentration of carbon dioxide (CO_2) in the atmosphere since 1700. (Figure courtesy of the Scripps Institution of Oceanography at U. C. San Diego)

climate known to have occurred since the pre-industrial era (1850–1900) is ultimately related to the approximately one degree Celsius increase in Earth's average yearly surface temperature that has taken place. This increase in temperature is attributed to elevated concentrations of carbon dioxide (CO_2) in the atmosphere. The current global average concentration of carbon dioxide in the atmosphere is approximately 420 ppm (0.04%), and this represents an increase of about 50% since the beginnings of the Industrial Revolution in the mid-1800s, when the global average concentration was about 280 ppm (Fig. 14.1). If nothing happens, the concentration of carbon dioxide will continue to increase over the coming decades. Even under conservative scenarios, future changes in climate are likely to include further increases in temperature with significant drying (drought) in some regions and increases in the frequency and severity of extreme weather events (IPCC 2023a, b).

Climate change can be manifested in two different ways. The first is the change in mean conditions (primarily with respect to temperature and moisture) and the second is a change in climate variability. In many instances, the latter is the more important. This has been demonstrated by the recent increases in the intensity and frequency of extreme weather events, which have resulted in major flooding and extensive and frequent wildfires. Although climate change is happening globally, the effects of climate change can differ locally, with some forest types or habitats more vulnerable than others.

14.2 Example from the Past

As mentioned earlier in this book, during the period of approximately 8000–5000 years ago (the Holocene Climatic Optimum), temperatures in the two regions were higher than at any other time during the Holocene Epoch. Evidence from fossil pollen assemblages indicates that these higher temperatures resulted in the displacement of red spruce upslope in the mountains of the Appalachians. This would happen again if global temperatures continue to increase. This upslope displacement took place only where the mountains were high enough. Where this was not the case, red spruce would have been extirpated. It has been hypothesized that some mountains in the Appalachians reached an elevation that permitted the survival of red spruce but were not high enough for Fraser fir, which generally occurs at higher elevation than red spruce, to do the same. It seems possible that this was the case for Whitetop Mountain in southwest Virginia. Only red spruce currently occurs on the summit of Whitetop Mountain (elevation 1682 m), but both red spruce and Fraser fir occur on nearby Mt. Rogers (1746 m), which is slightly higher.

14.3 Forest Ecosystems

There is no doubt that climate change is intimately entwined with the health and functioning of forest ecosystems. Some aspects of the impact of climate change will not be immediately apparent. Elevated temperatures and changes in moisture availability have the potential of affecting the geographic ranges of certain species of plants and animals as well as the timing of their life cycle events such as reproduction. For example, the fact that some plants are flowering earlier in the spring due to the warmer temperatures associated with climate change has been documented in a number of studies.

Distribution of tree species also can be affected by changing climatic conditions. The Climate Change Atlas describes current and projected geographical distribution of tree species under various scenarios (Peters et al. 2020). Projects are based on a General Circulation Model (GCM) and Representative Concentration Pathways (RCP) data. Using scenarios that represent the lower bounds of greenhouse gas (GHG) emission level scenarios, an examination of proposed distributions of some species such as red spruce indicates that they may be very restricted. Many Appalachian species may experience northward shifts, and Ozark species may become more restricted.

14.3.1 Forest Fires

Higher global temperatures will result in an increase in the number and intensity of forest fires, especially in forests located on xeric sites. An increase in fire intensity and occurrence has taken place, with changing temperature implicated. In the long-term, this will favor fire adapted trees such as pines and some oaks that are better adapted to fire-prone ecosystems. However, increased frequency of fire, particularly high intensity fires, may be detrimental regardless of adaptation. Moreover, species that are fire sensitive may be considerably diminished. When a forest fire occurs, most non-highly motile organisms are killed. This includes microorganisms in the soil. This results in post-fire communities that can be very different with respect to their composition and biodiversity.

14.3.2 Stress of Higher Temperatures

The stress of higher temperatures will lower the fitness of trees, making them more susceptible to fungal and insect pathogens. This will be devastating to some species of trees and likely will result in appreciable compositional changes in certain types of forests. Temperature and moisture are factors that influence trees and other plants throughout their life cycle. First, a certain level of fitness is necessary for a plant to divert the energy resources required for reproduction. Seed dispersal would seem largely unaffected by climate change except how it might impact vectors of animal-dispersed seeds. Seed germination and seedling survival are clearly affected by the existing conditions of temperature and moisture. Light availability could become important if the structure and composition of the surrounding vegetation has been altered by climate change. The latter also could affect growth performance, although temperature and moisture are much more important factors.

14.3.3 Elevated CO_2

Photosynthesis is the process by which green plants (and some other organisms) utilize solar energy to synthesize carbon-containing molecules from carbon dioxide and water. In green plants the process involves the green pigment chlorophyll and generates oxygen as a byproduct. To say that photosynthesis is the fundamental driving force for all life on the earth would be an understatement. Carbon dioxide is a relatively minor component (only 0.04%) of the Earth's atmosphere and can be a limiting factor for photosynthesis in some situations. The question might be asked as to what effects elevated levels have on photosynthesis and by default other plant processes. Studies have indicated that for some of the plants that have been investigated, elevated CO_2 increases net photosynthesis and decreases transpiration, thus affecting plant growth (Wand et al. 1999). Plant growth is manifestsed in increased leaf area, increased leaf thickness, and results in larger shoots, stems, and leaves (Pritchard et al. 1999).

14.3.4 Forest Mesophication

For more than half a century, forest ecologists have noted a general pattern of poor recruitment of species of oaks and hickories in the forests of both the Appalachians and the Ozarks (Abrams 1992; Olson et al. 2014; Rudolph et al. 2025). Forests in which representatives of these two taxa were once dominant have been reported to be characterized by an increasing importance of more mesophytic trees, particularly species of maple (especially red maple) and American beech. This change in composition has been attributed to an increased level of fire suppression and generally more moist conditions, both of which would create a more favorable environment for mesophytic species (Nowacki and Abrams 2008). Portions of the eastern United States are known to have become wetter over the past century, at least in part due to climate change (Ficklin et al. 2015). The effect of an anticipated increase in global temperatures is likely going to result in shifts in the distribution patterns of particular species of forest trees and by default compositional changes in forest composition. To what extent this will happen cannot be predicted. Fire suppression is likely to remain at its current level or even increase in the future. However, prescribed fires have been increased, and careful ignitions in systems where fire might represent an important form of disturbance and important form of restoration will be essential.

Other less well-defined factors also likely play some role in the poor recruitment of oaks and hickories. Many present-day forests were established during a period of widespread disturbance from logging, clear-cutting, and fire. Today, most of these same forests are being subjected to much less disturbance, and the successional patterns taking place might be expected to be different. This difference in succession is important in a number of ways. For example, the decline in species of oak and hickory has serious ecological consequences for wildlife (e.g., turkey [*Meleagris gallopavo*] and white-tailed deer [*Odocoileus virginianus*]) and equally serious economic consequences because of the reduction in wood resources, especially those derived from oak trees. All of what is mentioned above will have to factor into forest management in the future.

14.4 Subalpine Habitats

Subalpine habitats in the Appalachians are characterized by their generally cool and moist conditions. Many of the organisms that occur in these habitats require such conditions to survive. The species of plethodon salamanders associated with high-elevations forests represent one example. The Cheat Mountain salamander (*Plethodon nettingi*) is known to occur in a limited number of localities in the mountains of West Virginia and Virginia, and populations of this animal are already known to be declining (Fig. 14.2). If elevated temperatures cause subalpine forests to decrease in extent, the populations of organisms associated with these forests will decrease in size, with some organisms likely to be extirpated from particular localities. Because of their already limited aerial extent and the rather severe environmental conditions to which they are subjected (e.g., high levels of ultraviolet radiation, low winter temperatures, and usually thin, rocky soils), any species that exists in low numbers is always susceptible to being eliminated with habitat changes.

Bunchberry or dwarf cornel (*Cornus canadensis*) is a slow-growing herbaceous perennial plant found throughout the northern United States and Canada (Fig. 14.3). The plant is only about 10 to 20 cm tall, and the above ground stems arise from an underground rhizome. As a member of the same genus as flowering dogwood

Fig. 14.2 The Cheat Mountain salamander (*Plethodon nettingi*), which has a limited distribution in the Appalachians. (Photo by Brady O'Brien)

Fig. 14.3 Bunchberry (*Cornus canadensis*), a plant known from only a few scattered localities at high elevations in the Appalachians. (Photo by Steven L. Stephenson)

(*Cornus florida*), bunchberry has similar flowers, but with only one flower on each plant. In the Appalachians, bunchberry is known from only a few scattered localities at high elevations in Virginia and West Virginia (Strausbaugh and Core 1978). In fact, one population occurs near the summit of Spruce Knob (elevation 1482 m) in West Virginia. Bunchberry is associated with cool, moist microhabitats (Wherry 1934). Increased global temperatures would result in a reduction in the extent of such habitats, and this species is likely to be extirpated from the localities mentioned above. Obviously, this would have little effect on the species as a whole because of its relative abundance at higher latitudes. However, the loss of even a single species from a portion of its range cannot be discounted, and bunchberry is not alone in having a restricted distribution in the portion of the Appalachian system considered in this book. Similarly, the Ozarks is a region of high endemism and therefore many species there will be vulnerable to climate changes.

Fig. 14.4 Encroachment of the surrounding forest into Cranberry Glades in the Appalachians of West Virginia. (Photo by Steven L. Stephenson)

14.5 Wetland Habitats

Higher global temperatures and the resultant more xeric conditions will result in a decline in wetland areas, reducing a habitat that is essential for the survival of organisms ranging from insects to birds. This is especially true for high-elevation bogs, where some of the organisms present are disjuncts from higher latitudes or are regionally or globally rare. An encroachment of the surrounding vegetation upon bogs is already occurring at places such as Cranberry Glades in the mountains of West Virginia and is likely to be accelerated by climate change (Fig. 14.4). Edens (1972) reported that one of the five glades that collectively made up Cranberry Glades when it was surveyed in 1911 was overgrown 60 years later. Wetlands such as bogs already suffer greatly as a result of human activities, which only compounds the difficulties faced by the organisms found in these habitats.

14.6 Microbes

14.6.1 The Microbial Loop

Terrestrial soil microbes—primarily bacteria, fungi, and eumycetozoans—play a critical role in mediating biogeochemical cycles. However, their response to elevated temperatures and lower moisture levels is unknown. Eumycetozoans (commonly called slime molds) are abundant in forest soils (Stephenson et al. 2011) and along with nematodes are involved in what is known as the "microbial loop" (Raynaud et al. 2006). In brief, exudates from plant roots (including those of trees) introduce nutrients into the soil. These nutrients are then taken up by soil bacteria, resulting in an increase in bacterial biomass. If it were not for the presence of bacteria-consuming organisms in the soil, nutrients would be locked up in this bacterial biomass and thus no longer available for use at higher trophic levels. Although not generally appreciated, eumycetozoans are among the most important of the organisms that feed upon bacteria. The distribution of eumycetozoans (and also nematodes) in the soil is influenced by a number of factors, with moisture, soil nutrient status, and the amount of organic matter the most important. All of these can be affected directly (moisture) or indirectly (nutrient status and organic matter) by climate change. This interaction that occurs in the soil has both a "top-down effect" on the control of bacteria numbers as well as a "bottom-up effect" on the nutrient-acquiring microorganisms involved.

14.6.2 Mycorrhizal Fungi

The importance of fungi to forests ecosystems cannot be overstated, as was described in Chap. 9. Fungi are decomposers of all types of organic matter, and some are parasites or pathogens. The story of the chestnut blight fungus and American chestnut (Chap. 13) provides an example of just how much a fungus can affect a forest ecosystem in the Appalachians. As noted in Chap. 9, all trees (except for unusual examples such as palms) are dependent upon mycorrhizal fungi. Assessing the possible effects of climate change on fungi is difficult, especially in field-based studies. Presumably, any reduction in the fitness of a tree (or any other plant) would affect the associated mycorrhizal fungus, since the latter derives its nutrients from the tree. This represents a biological system in forests that clearly warrants additional study.

14.6.3 Decomposers

Decomposition of wood, litter, and other types of plant debris by fungi is clearly affected by the level of moisture present, with temperature playing a lesser role. When temperatures are low during winter, little decomposition takes place. The same is true during summer when the substrates being decomposed are very dry, but in general warmer temperatures enhance decomposition rates. However, the changes brought on by climate change would be very minor except in one situation. If the elevated temperatures of climate change result in an increase in fire frequency, the fungi involved in decomposition would be severely affected. An intense fire would likely kill most of the fungi present or, at the very least, reduce fungal diversity. This was documented by Alanbagi et al. (2021) in a study carried out in the forests of the Ozarks in northwest Arkansas.

14.6.4 Parasites and Pathogens

Invasive species of parasitic and pathogenic fungi are among the most serious threats to forests, and their cryptic nature and dispersal ability, along with often poor detection methods, allow for their rapid and extensive invasion. Most examples were initially introduced to North America (and consequently to the Appalachians and Ozarks) when their hosts were brought here, either by design or unintentionally. The relationship involving a parasitic/pathogenic fungus can be considered as a three-way interaction among the fungus, the host/substrate, and the environment (Alexander 2010). Obviously, the latter is affected by climate change. Less cryptic are damaging insect pathogens, although they also a serious threat to forests. Like the fungi noted above, many of these were inadvertently introduced into North America. However, there are certain indigenous species that sometimes undergo massive population increases, often with devastating effects in local areas. Insect populations, with respect to both abundance and distribution, are strongly controlled by climatic conditions.

Climate change can directly affect populations of pathogens and parasites by limiting (or enhancing) sporulation and influencing the likelihood of successful infection of their hosts. Many species are very sensitive to the timing and quantity of precipitation, the ambient temperature, relative humidity, and various factors (e.g., moisture or temperature stress) relating to the host plant (Wood et al. 2018). It is well established that some species of already established pathogens/parasites suddenly become more destructive or expand into new localities and hosts. Determining why this happens is often difficult or impossible, and understanding the changes associated with climate change will make it even more so.

Herbivores also can impact the species composition, ecosystem function, and socioeconomic value of forests. Both herbivores and pathogens are an integral part of forests but sometimes produce undesirable effects and result in the degradation of forest resources. For the entire United States, just the activities of just a few of the more important species of forest fungi and insects have an economic cost that probably exceed one billion dollars a year. Climate change could alter patterns of disturbance from insect herbivores if changes occur in the abundance of their natural enemies. Because many insects have a short life cycle, some have a limited period of mobility, and all possess some degree of physiological sensitivity to temperature, even modest climate change can impact their reproductive potential and thus their abundance and distribution (Ayres and Lombardero 2000).

14.7 Invasive Plants

Invasive species tend to be characterized by high dispersal rates, rapid growth rates with short generation times, and the capacity to tolerate a broad range of environmental conditions. Collectively, these traits greatly enhance their ability to cope with rapid changes in abiotic and biotic conditions such as those associated with climate change. Invasive plants are frequently associated with sites that have been disturbed but also can colonize undisturbed areas. There is abundant literature on the ecological effects of invasive plants. Many of these are related to the ability of non-native plants to outcompete natives, to capitalize on resources, and to reduce biological diversity, particularly in the shrub and herbaceous layers of a forest. Considerable disturbance (e.g., the loss of the cover of vegetation at some sites) would result from extreme weather events, and disturbance also could result from the more subtle effects of climate change on the structure and composition of the vegetation at a particular site (e.g., loss canopy trees as a result of moisture stress when moisture conditions become more xeric over time).

Any of the disturbances noted above could allow nonnative invasive plants to expand within the habitats where they already occur and also to become established in new habitats. As growing seasons become longer, the likelihood of species now confined to warm climates to migrate northward will increase. Moreover, the warmer conditions also enhance the chances of warm-weather species to become newly established in regions where they were formerly absent. Native plants also will be affected by climate change, and some have the potential to become invasive when environmental conditions allow them to shift their geographic ranges into novel habitats (Finch et al. 2021). The same situation applies to fungi and animals, with pathogenic/parasitic fungi and insects the two groups of most concern. The increased frequency, intensity, and duration of extreme weather events associated with climate change can affect invasive species (insects, plants, and fungi) in a number of ways. Strong winds such as those of hurricanes can carry propagules (e.g., the seeds of plants and the spores of fungi) and reduce existing vegetation or create bare soil. All of these would enhance opportunities for the establishment of invasives at new localities. More frequent and more expansive flooding resulting from climate change also would likely contribute to the spread of invasives.

14.8 Carbon Sinks

Globally, forests store large amounts of carbon sequestered from the atmosphere and retained in living and dead biomass and soil. As such, they serve as carbon sinks. By definition, a carbon sink sequesters (through photosynthesis) more carbon (as carbon dioxide) from the atmosphere than the carbon they release (Whitehead 2011). On average, trees sequester around twice as much carbon dioxide as they release. As such, trees (and the forests they make up) are an essential means of helping fight climate change. Any major reduction in total extent of the forests of Appalachians and the Ozarks as a result of climate change would have serious consequences.

Although elevated CO_2 would theoretically increase net photosynthesis and thus biomass in a forest, other factors related to climate change are likely to have a more substantial impact. Long-term changes in composition and structure resulting from a shift in mean conditions of temperature and moisture (e.g., a particular site becoming more xeric) would certainly be important. On a short-term basis, outbreaks of insect and fungal pathogens have the potential of bringing about major forest-level reductions in carbon storage. These pathogens can reduce tree growth, kill trees, and alter the distribution of carbon (i.e., from living biomass to dead biomass) within a forest (Cobb et al. 2012). An increase in dead biomass as represented by woody debris also can lead to changes in the susceptibility of a forest to fire or the behavior of fire if the latter occurs. Large scale fires can quickly reduce carbon storage to carbon dioxide in the atmosphere.

Trees accumulate biomass and the amount of carbon they sequester as they grow. The species of tree involved and the structure of the forest community in which the tree occurs determine the rate at which sequestration takes place. For naturally seeded understory tolerant species, establishment after widespread disturbance can result in a forest community characterized by a high density of trees of approximately the same age and size (Fig. 14.5). In forest communities where the trees are older and less understory tolerant, the structure is very different, with a lower density and much larger trees (Fig. 14.6). However, it is possible that the two communities might have comparable amounts of stored carbon.

Fig. 14.5 A self-seeded red spruce forest with a high density of trees. (Photo by Steven L. Stephenson)

Fig. 14.6 A more mature forest in the Appalachians with a lower density of trees. (Photo by Steven L. Stephenson)

Considerable areas of forests are under private ownership in both the Appalachians and Ozarks. For the entire United States, about 58% of all forested land is privately owned.

Various carbon offset programs have been developed to credit landowners for managing forests to maintain and increase long-term carbon storage. These programs focus on the carbon sinks represented both by live trees and standing dead trees, and the goal is to maintain the long-term stability of forest carbon. Unfortunately, the huge scale of the overall goal of increasing carbon storage limits the practicality these efforts.

In summary, although it is not possible to predict the exact effects of climate change, there is no doubt that the climate to which we have been accustomed is no longer what we can expect in the future. This must be a consideration when developing both short-term and long-term management plans for forests and the organisms that inhabit them.

References

Abrams MD (1992) Fire and the development of oak forests. Bioscience 42:346–353

Alanbagi RA, Ashuwzili FF, Stephenson SL (2021) Prescribed burning reduces species richness of wood-decay fungi in the forests of Northwest Arkansas. Saudi J Biol Sci 28:936–941

Alexander HM (2010) Disease in natural plant populations, communities, and ecosystems: insights into ecological and evolutionary processes. Plant Dis 94:492–503

Ayres MP, Lombardero MJ (2000) Assessing the consequences of global change for forest disturbance from herbivores and pathogens. Sci Total Environ 262:262–286

Cobb RC, Chan MN, Meentemeyer RK, Rizzo DM (2012) Common factors drive disease and coarse woody debris dynamics in forest impacted by sudden oak death. Ecosystems 15:242–255

Edens DL (1972) A correction in the naming of the cranberry glades in West Virginia. Castanea 37:229–234

Ficklin DL, Maxwell JT, Letsinger SL, Gholizadeh H (2015) A climatic deconstruction of recent drought trends in the United States. Environ Res Lett 4:044009

Finch DM, Butler JL, Runyon JB et al (2021) Effects of climate change on invasive species. In: Poland TM, Patel-Weynand, Finch DM et al (eds) Invasive species in forests and rangelands of the United States. Springer Nature, Gewerbestrase, pp 57–84

Intergovernmental Panel on Climate Change (IPCC) (2023a) Climate change 2023: the physical science basis. In: Sixth Assessment Report of the Intergovernmental Panel on Climate Change. Cambridge University Press, Cambridge, p 996

Intergovernmental Panel on Climate Change (IPCC) (2023b) Climate change 2023: synthesis report. In: Core Writing Team, H HL, Romero J (eds) Contribution of Working Groups I, II and III to the Sixth Assessment Report of the Intergovernmental Panel on Climate Change. IPCC, Geneva, pp 35–115. https://doi.org/10.59327/IPCC/AR6-9789291691647

Nowacki GJ, Abrams MD (2008) The demise of fire and "mesophication" of forests in the eastern United States. Bioscience 58:123–138

Olson MG, Stevenson AP, Knapp BO, Kabrick JM, Jessen RG (2014) Is there evidence of mesophication of oak forests in the Missouri Ozarks? In: Groninger JW, Holzmueller EJ, Nielson CK, Dey DC (eds) Proceedings, 19th Central Hardwood Forest Conference, March 10–12, 2014, Carbondale Illinois, General Technical Report NRS-P-14. United States Department of Agriculture, Forest Service, Northern Research Station, Newtown Square, PA

Parmesan C (2006) Ecological and evolutionary responses to recent climate change. Annu Rev Ecol Evol Syst 37:637–669

Peters MP, Prasad AM, Matthews SN, Iverson LR (2020) Climate change tree atlas, version 4. United States Forest Service. Northern Research Station and Northern Institute of Applied Climate Science, Delaware, OH. https://www.nrs.fs.fed.us/atlas

Pritchard SG, Rogers HH, Prior SA, Peterson CM (1999) Elevated CO_2 and plant structure: a review. Glob Chang Biol 5:807–837

Raynaud X, Lata J, Leadley P (2006) Soil microbial loop and nutrient uptake by plants: a test using a coupled C:N model of plant-microbial interactions. Plant Soil 287:95–116

Rudolph AJ, Snell RS, McCarthy BC (2025) A decades-long case study: understanding the effects of mesophication on the forest community with emphasis on *Carya* spp. dynamics. J Torrey Bot Soc 152:10–22

Stephenson SL, Fiore-Donno A-M, Schnittler M (2011) Myxomycetes in soil. Soil Biol Biochem 43:2237–2242

Strausbaugh PD, Core EL (1978) Flora of West Virginia. Seneca Books, Morgantown, WV

Wand SJE, Midgley GF, Jones MH, Curtis PS (1999) Responses of wild C4 and C3 grass (Poaceae) species to elevated atmospheric CO_2 concentration: a meta-analytic test of current theories and perceptions. Glob Chang Biol 5:723–741

Wherry ET (1934) Temperature relations of the bunchberry, Cornus canadensis. Ecology 15:440–443

Whitehead D (2011) Forests as carbon sinks—benefits and consequences. Tree Physiol 31:893–902

Wood JD, Knapp BO, Muzika RM, Stambaugh MC, Gu L (2018) The importance of drought-pathogen interactions in driving oak mortality events in the Ozark Border Region. Environ Res Lett 13(1):015004

Appendixes

Appendix A: Common and Taxonomic Names Used in This Book

Common name	Taxonomic name
Plants	
Allegheny chinquapin	*Castanea pumila*
Alum root	*Heuchera parviflora*
American beech	*Fagus grandifolia*
American chestnut	*Castanea dentata*
American holly	*Ilex opaca*
American spikenard	*Aralia racemosa*
American wintergreen	*Pyrola americana*
American witch-hazel	*Hamamelis virginiana*
American yew	*Taxus canadensis*
Apple	*Malus* spp.
Arkansas alumroot	*Heuchera villosa* var. *arkansana*
Ashe's juniper	*Juniperus ashei*
Asian ginseng	*Panax ginseng*
Balsam fir	*Abies balsamea*
Basswood	*Tilia americana* var. *heterophylla*
Beaked sedge	*Carex rostrata*
Bee balm	*Monarda didyma*
Bentley's coralroot	*Corallorhiza bentleyi*
Big bluestem	*Andropogon gerardii*
Big-flowered coreopsis	*Coreopsis grandiflora*
Bird's foot violet	*Viola pedata*

S. L. Stephenson, R.-M. Muzika, *Forests of the Southern Appalachian Highlands and Interior Highlands of the United States*, Plant and Vegetation 20,
https://doi.org/10.1007/978-3-032-01155-8

Common name	Taxonomic name
Birdsfoot trefoil	*Lotus corniculatus*
Bitternut hickory	*Carya cordiformis*
Black birch	*Betula lenta*
Black cherry	*Prunus serotina*
Black cohosh	*Actaea racemosa*
Black gum	*Nyssa sylvatica*
Black hickory (or Texas hickory)	*Carya texana*
Black locust	*Robinia pseudoacacia*
Black oak	*Quercus velutina*
Black spruce	*Picea mariana*
Black walnut	*Juglans nigra*
Black-grass rush	*Juncus gerardii*
Blackjack oak	*Quercus marilandica*
Blazing star	*Liatris squarrosa*
Bloodroot	*Sanguinaria canadensis*
Blue ash	*Fraxinus quadrangulata*
Blue Ridge ragwort	*Packera millefolium*
Bog goldenrod	*Solidago uliginosa*
Bog willow-herb	*Epilobium leptophyllum*
Bordered mnium	*Mnium marginatum*
Box elder	*Acer negundo*
Bracken fern	*Pteridium aquilinium*
Bradley's spleenwort	*Asplenium bradleyi*
Broad-leaved cattail	*Typha latifolia*
Broomsedge	*Andropogon virginicus*
Bulblet bladder fern	*Cystopteris bulbifera*
Burdock	*Arctium minus*
Bush honeysuckle	*Lonicera maackii*
Bush's purple coneflower	*Echinacea paradoxa*
Bush's skullcap	*Scutellaria bushii*
Butterfly weed	*Asclepius tuberosa*
Button snakeroot	*Eryngium yuccifolium*
Canada mayflower	*Maianthemum canadense*
Canada violet	*Viola canadensis*
Canadian burnet	*Sanguisorba canadensis*
Canadian hemlock (or hemlock)	*Tsuga canadensis*
Canadian moonseed	*Menispermum canadense*
Canadian summer bluet	*Houstonia canadensis*
Carolina ash	*Fraxinus carolina*
Carolina hemlock	*Tsuga caroliniana*
Cat greenbrier	*Smilax glauca*
Catawba rosebay	*Rhododendron catawbiense*
Celestial lily	*Nemastylis nutalli*
Chestnut oak	*Quercus montana*

Common name	Taxonomic name
Chinquapin oak	*Quercus muehlenbergii*
Christmas fern	*Polystichum acrostichoides*
Churchmouse threeawn	*Aristida dichotoma*
Cinnamon fern	*Osmundastrum cinnamomeum*
Coltsfoot	*Tussilago farfara*
Common blue violet	*Viola sororia*
Common clubmoss	*Lycopodium clavatum*
Common evening primrose	*Oenothera biennis*
Common greenbrier	*Smilax rotundifolia*
Common ironweed	*Veronia fasciculata*
Common sundew	*Drosera rotundifolia*
Cotton grass	*Eriophorum virginicum*
Crown-vetch	*Secuorigea varia*
Cucumber magnolia	*Magnolia acuminata*
Dandelion	*Taraxacum officinale*
Deerberry	*Vaccinium stamineum*
Delicate fern moss	*Thuidium delicatulum*
Dewberry	*Rubus hispidus*
Downy serviceberry	*Amelanchier arborea*
Drummond's St. John's-wort	*Hypericum drummondii*
Duck potato	*Sagidttaria laatifolia*
Dutchman's breeches	*Dicentra cucullaria*
Dwarf cinquefoil	*Potentilla canadensis*
Dwarf ginseng	*Panax trifolius*
Dwarf larkspur	*Delphinium tricorne*
Dwarf spiderwort	*Tradescantia longipes*
Dwarf sundew	*Drosera brevifolia*
Early ladies' tresses	*Spiranthes lucida*
Eastern cottonwood	*Populus deltoides*
Eastern prickly pear	*Opuntia humifusa*
Eastern redbud (or redbud)	*Cercis canadensis*
Eastern white pine	*Pinus strobus*
Ebony spleenwort	*Asplenium platyneuron*
Evergreen shield fern	*Dryopteris intermedia*
False Solomon's seal	*Polygonatum bifloum*
Farkleberry	*Vaccinium arboreum*
Fire cherry	*Prunus pensylvanica*
Flame azalea	*Rhododendron calendulaceum*
Flat-topped white aster	*Aster umbellatus*
Flowering dogwood	*Cornus florida*
Flowering spurge	*Euphorbia corollata*
Forest goldenrod	*Solidago arguta*
Forked panic grass	*Dichanthelium dichotomum*
Fox grape	*Vitis vulpina*

Common name	Taxonomic name
Fraser fir	*Abies fraseri*
Frasier magnolia	*Magnolia fraseri*
Garlic mustard	*Alliaria petiolata*
Ghost pipe	*Monotropa uniflora*
Ginseng	*Panax quinquefolius*
Goatsbeard	*Aruncus dioicus*
Goldenrods	*Solidago* spp.
Goldenseal	*Hydrastis canadensis*
Goosefoot	*Chenopodium berlandieri*
Grass pink	*Calopogon tuberosus*
Grass-of-parnassus	*Parnassia grandifolia*
Great laurel	*Rhododendron maximum*
Greater purple-fringed orchid	*Platanthera grandiflora*
Green ash	*Fraxinus pennsylvanica*
Green trillium	*Trillium viridescens*
Ground nut	*Apios americana*
Gum bumelia	*Bumelia lanuginosa*
Haircap moss	*Polytrichum* spp.
Hairy disporum	*Prosartes lanuginosa*
Hairy wood mint	*Blephilia ciliata*
Hawthorn	*Crataegus* spp.
Hay-scented fern	*Dennstaedtia punctilobula*
Hazel alder	*Alnus serrulata*
Hollyhock	*Alcea rosea*
Hophornbeam	*Ostrya* spp.
Hornbeam	*Carpinus* spp.
Horsebalm	*Collinsonia canadensis*
Jack pine	*Pinus banksiana*
Jack-in-the-pulpit	*Arisaema triphyllum*
Japanese honeysuckle	*Lonicera japonica*
Japanese stiltgrass	*Microstegium vimineum*
Jerusaleum artichoke	*Helianthus tuberosa*
Kate's Mountain clover	*Trifolium virginicum*
Kudzu	*Pueraria* montana
Laceflower	*Ptilimnium nuttallii*
Lamb's quarters	*Chenopodium album*
Lance-leaf ragweed	*Ambrosia bidentata*
Large cranberry	*Vaccinium macrocarpon*
Large flower bellwort	*Uvularia grandiflora*
Large-flowered white trillium	*Trillium grandiflorum*
Late purple aster	*Symphyotrichum patens*
Leafy bulrush	*Scirpus polyphyllus*
Leptohymenium moss	*Leptohymenium sharpii*
Little bluestem	*Schizachyrium scoparium*

Common name	Taxonomic name
Littleflower alumroot	*Heuchera parviflora*
Liverleaf	*Hepatica* spp.
Lobed spleenwort	*Asplenium pinnatifidum*
Loblolly pine	*Pinus taeda*
Long-stalked holly	*Ilex collina*
Low-bush blueberry	*Vaccinium pallidum*
Maidenhair fern	*Adiantum pedatum*
Maple-leaf oak	*Quercus acerifolia*
Mayapple	*Podophyllum peltatum*
Mead's milkweed	*Asclepias meadii*
Meadow garlic	*Allium canadense*
Mexican plum	*Prunus mexicana*
Millboro leatherflower	*Clematis viticaulis*
Mimosa	*Albizia julibrissin*
Minnie-bush	*Rhododendron pilosum*
Missouri bladderpod	*Lesquerella filiformis*
Missouri primrose	*Oenothera speciosa*
Mockernut hickory	*Carya alba*
Morning glory	*Ipomea* spp.
Moss phlox	*Phlox subulata*
Moss with no common name	*Calliergonella curvifolia*
Moss with no common name	*Donrichardsia pringlei*
Moss with no common name	*Ptychomitrium sinense*
Moss with no common name	*Racomitrium aduncoides*
Moss with no common name	*Venturiella sinensis*
Mountain ash	*Sorbus americana*
Mountain cranberry	*Vaccinium erythrocarpum*
Mountain fern moss	*Hylocomium splendens*
Mountain holly	*Ilex montana*
Mountain laurel	*Kalmia latifolia*
Mountain maple	*Acer spicatum*
Mountain nailwort	*Paronychia montana*
Mountain oatgrass	*Danthonia compressa*
Mountain pimpernel	*Taenidia montana*
Mulberry	*Morus ruba*
Multiflora rose	*Rosa multiflora*
Muscadine	*Vitis rotundifolia*
Naked-stem sunflower	*Helianthus occidentalis*
Narrowleaf mountain mint	*Pycnanthemum tenuifolium*
Newton's larkspur	*Delphinium newtonianum*
Northern red oak	*Quercus rubra*
Northern white cedar	*Thuja occid*entalis
Ohio buckeye	*Aesculus glabra*
One-seed croton	*Croton willdenowii*

Common name	Taxonomic name
Ozark calamint	*Clinopodium arkansanum*
Ozark chinquapin	*Castanea ozarkensis*
Ozark cornsalad	*Valerianella ozarkana*
Ozark mountain goldenrod	*Solidago ouachitensis*
Ozark spiderwort	*Tradescantia ozarkana*
Ozark trillium	*Trillium ozarkanum*
Ozark witch-hazel	*Hamamelis vernalis*
Ozarks wild crocus	*Tradescantia longipes*
Painted trillium	*Trillium undulatum*
Paper birch	*Betula cordifolia* and *B. papyrifera*
Pawpaw	*Asimina triloba*
Peat moss	*Sphagnum* spp.
Pelton's rose-gentian	*Sabatia arkansana*
Pennsylvania sedge	*Carex pensylvanica*
Perennial ryegrass	*Lolium cneata*
Persimmon	*Diospyros virginiana*
Peter's Mountain mallow	*Iliamna corei*
Pignut hickory	*Carya glabra*
Pincushion moss	*Leucobryum albidum*
Pitch pine	*Pinus rigida*
Poison hemlock	*Conium maculatum*
Poison ivy	*Toxicodendron radicans*
Pokeweed	*Phytolacca americana*
Post oak	*Quercus stellata*
Poverty grass	*Aristida dichotoma*
Poverty oat grass	*Danthonia spicata*
Prairie dock	*Silphium integrifolium*
Prairie sunflower	*Helianthus pauciflorus*
Prairie violet	*Viola pedatifida*
Princess tree	*Paulownia tomentosa*
Pumpkin ash	*Fraxinus profunda*
Purple pitcher plant	*Sarracenia purpurea*
Purple trillium	*Trillium recurvatum*
Purple-fringed orchid	*Habenaria fimbriata*
Ramp	*Allium tricoccum*
Red cedar (or eastern red cedar)	*Juniperus virginiana*
Red clover	*Trifolium pratense*
Red hickory	*Carya ovalis*
Red maple	*Acer rubrum*
Red pine	*Pinus resinosa*
Red trillium	*Trillium erectum*
Redbay	*Persea borbonia*
Rein orchid	*Platanthera flave*
Riddell's goldernrod	*Oligoneuron riddellii*

Common name	Taxonomic name
Rose pogonia	*Pogonia ophioglossoides*
Rosinweed	*Silphium intergrifolium*
Roundleaf catchfly	*Silene rotundifolia*
Roundleaf sundew	*Drosera rotundifolia*
Roundleaf violet	*Viola rotundifolia*
Running cedar	*Diphasiastrum digitatum*
Saltmarsh bulrush	*Bolboschoenus robustus*
Sassafras	*Sassafras albidum*
Scarlet oak	*Quercus coccinea*
Sensitive fern	*Onocleas sensibilis*
Sericea lespedeza	*Lespedeza cuneata*
Service berry	*Amelanchier arborea*
Shagbark hickory	*Carya ovata*
Shale barren buckwheat	*Eriogonum allenii*
Shale-barren evening primrose	*Oenothera argillicola*
Shale-barren ragwort	*Packera antennariifolia*
Shale-barren rockcress	*Arabis serotina*
Short-tailed rush	*Juncus subcaudatus*
Shortleaf pine	*Pinus echinata*
Showy lady's slipper orchid	*Cypripedium reginae*
Shumard oak	*Quercus shumardii*
Side oats grama grass	*Bouteloua curtipendula*
Silverbell	*Halersia monticola*
Skunk cabbage	*Symplocarpus foetidus*
Slender mountain-mint	*Pycnanthemum tenuifolium*
Slender spikerush	*Eleocharis tenuis*
Slippery elm	*Ulmus rubra*
Small cranberry	*Vaccinium oxycoccos*
Smooth alder	*Alnus serrulata*
Smooth sunflower	*Helianthus laevigatus*
Sourwood	*Oxydendrum arboretum*
Southern red oak	*Quercus falcata*
Southern maidenhair fern	*Adiantum capillus-veneris*
Spring beauty	*Claytonia virginica*
St. John's wort	*Hypericum perforatum*
Staghorn sumac	*Rhus typhina*
Stiff clubmoss	*Spinulum annotinum*
Striped maple	*Acer pensylvanicum*
Striped prince's pine	*Chimaphila maculata*
Sugar maple	*Acer saccharum*
Summer grape	*Vitis aestivalis*
Sunflowers	*Helianthus* spp.
Swamp lousewort	*Pedicularis lanceolata*
Switch grass	*Panicum virgatum*

Common name	Taxonomic name
Sycamore	*Platanus occidentalis*
Table Mountain pine	*Pinus pungens*
Tall fescue	*Festuca arundinacea*
Tall larkspur	*Delphinium exaltatum*
Tatarian honeyhuckle	*Lonicera tatarica*
Teaberry	*Gaultheria procumbens*
Threadleaf evening primrose	*Oenothera linifolia*
Three-lobed bazzania	*Bazzania trilobata*
Three-part violet	*Viola tripartita*
Three-seeded sedge	*Carex trisperma*
Toadshade trillium	*Trillium sessile*
Toothworts	*Dentaria* spp.
Trailing arbutus	*Epigaea repens*
Tree clubmoss	*Dendrolycopodium obscurum*
Tree of heaven	*Ailanthus altissima*
Trelease's larkspur	*Delphinium treleasei*
True asters	*Aster* spp.
Turk's cap lily	*Lilium superbum*
Virginia creeper	*Parthenocissus quinquefolia*
Virginia mountain-mint	*Pycnanthemum virginianum*
Virginia pine	*Pinus virginiana*
Virginia round-leaf birch	*Betula uber*
Walking fern	*Asplenium rhizophyllum*
Water hemlock	*Cicuta maculata*
White ash	*Fraxinus americana*
White baneberry	*Actaea pachypoda*
White lady's slipper orchid	*Cypripedium candidum*
White oak	*Quercus alba*
White pine	*Pinus strobus*
White snakeroot	*Ageratina altissima*
White spruce	*Picea glauca*
White trillium	*Trillium grandiflorum*
White walnut	*Juglans cinerea*
White-haired clematis	*Clematis albicoma*
Whorled milkweed	*Asclepias verticillata*
Whorled wood aster	*Oclemena acuminata*
Wild bergamot	*Monarda fistulosa*
Wild ginger	*Asarum canadense*
Wild raisin	*Viburnum cassinoides*
Wild sarsaparilla	*Aralia nudicaulis*
Wild strawberry	*Fragaria virginiana*
Wild sweet William	*Phlox maculata* ssp. *pyramidalis*
Willow	*Salix* spp.
Windblown moss	*Dicranum scoparium*

Common name	Taxonomic name
Woodland rush	*Juncus subcaudatus*
Woodland sunflower	*Heliantus divaricatus*
Yellow birch	*Betula alleghaniensis*
Yellow buckeye	*Aesculus flava*
Yellow coneflower	*Echinacea paradoxa*
Yellow fringed orchid	*Platanthera ciliaris*
Yellow loosestrife	*Lysimachia quadriflora*
Yellow pine	*Pinus echinata*
Yellow poplar	*Liridodendron tulipifera*
Yellow rocket	*Barbarea vulgaris*
Yellow trout lily (Appalachians)	*Erythronium americanum*
Yellow trout lily (Ozarks)	*Erythronium rostratum*
Animals	
Allegheny woodrat	*Neotoma magister*
American bison	*Bison bison*
American crow	*Corvus brachyrhynchos*
American giant millipede	*Narceus americanus*
American kestrel	*Falco sparverius*
American mink	*Neogale vison*
American pygmy shrew	*Sorex hoyi*
American redstart	*Setophaga ruticilla*
American robin	*Turdus migratorius*
American toad	*Anaxyrus americanus*
Ancient bison	*Bison antiquus*
Ancient horse	*Equus scotti*
Appalachian cottontail	*Sylvilagus obscurus*
Balsam woolly adelgid	*Adelges piceae*
Barred owl	*Strix varia*
Beaver	*Castor canadensis*
Bed bug	*Cimex lectularius* and *C. hemipterus*
Beech leaf disease nematode	*Ditylenchus crenatae* ssp. *mccanna*
Big brown bat	*Eptesicus fucus*
Black and yellow garden spider	*Argiope aurantia*
Black bear	*Ursus americanus*
Black racer	*Coluber constrictor*
Black rat snake	*Pantherophis obsoletus*
Black-and-white warbler	*Mniotilta varia*
Blackburnian warbler	*Setophaga fusca*
Blue jay	*Cyanocitta cristata*
Blue-gray gnatcatcher	*Polioptila caerulea*
Bobcat	*Lynx rufus*
Bog copper	*Lycaena epixanthe*
Brown recluse	*Loxosceles reclusa*
Buffalo	*Bison bison*

Common name	Taxonomic name
Bumblebee moth	Hemaris *spp.*
Caribou	*Rangifer tarandus*
Carolina mantis	*Stagmomantis carolina*
Cecropia moth	*Hyalophora cecropia*
Cerulean warbler	*Dendroicea curulea*
Cheat Mountain salamander	*Plethodon nettingi*
Cheat threetooth snail	*Triodopsis platysayoides*
Chestnut-sided warbler	*Setophaga pensylvanica*
Chinese mantis	*Tenodera sinensis*
Chipmunk	*Tamias striatus*
Clench's middle-toothed land snail	*Mesodon clenchi*
Common yellowthroat	*Geothlypis trichas*
Copperhead	*Agkistrodon contortrix*
Coyote	*Canis latrans*
Deer mouse	*Peromyscus maniculatus*
Dire wolf	*Aenocyon dirus*
Dixie caverns salamander	*Plethodon dixi*
Eastern box turtle	*Terrapene carolina*
Eastern collared lizard	*Crotaphytus collaris*
Eastern cottontail	*Sylvilagus floridanus*
Eastern deer mouse	*Peromyscus maniculatus*
Eastern elk	*Cervus canadensis*
Eastern fence lizard	*Scelopocus undulatus*
Eastern garter snake	*Thamnophis sirtalis*
Eastern *giant swallowtail*	*Papilio cresphontes*
Eastern hellbender	*Cryptobranchus alleganiensis* ssp. *alleganiensis*
Eastern Hercules beetle	*Dynastes tityus*
Eastern newt	*Nogtophthalmus viridescens*
Eastern red bat	*Lasiurus borealis*
Eastern red-backed vole	*Clethrionomus gapperi*
Eastern screech owl	*Megascops asio*
Eastern spotted skunk	*Spilogale putorius*
Eastern tent caterpillar	*Malacosoma Americanum*
Eastern tiger swallowtail	*Palilio glaucus*
Eastern woodrat	*Neotoma floridana*
Eastern worm snake	*Carphophis amoenus*
Emerald ash borer	*Agrilus planipennis*
European mantis	*Mantis religiosa*
European starling	*Sturmus vugari*
Fall cankerworm	*Alsophila pometaria*
Fire ant	*Solenopsis invicta*
Five-lined skink	*Plestiodon fasciatus*
Four-toed salamander	*Hemidactylium scutatum*

Common name	Taxonomic name
Giant beaver	*Castoroides ohioensis*
Giant bison	*Bison latifrons*
Giant darner	*Anax walsinghami*
Giant ground sloth	*Magalonyx jeffersonii*
Glyptodont	*Glyptodon* sp.
Goldenrod cellophane bee	*Colletes solidaginis*
Gray fox	*Urocyon cinereoargenteus*
Gray squirrel	*Sciurus carolinenosis*
Gray tree frog	*Dryophytes versicolor*
Green frog	*Lithobates clamitans*
Ground beetles	*Calosoma* spp.
Ground hog	*Marmota monax*
Half-wing geometer	*Phigalia titea*
Harris' checkerspot	*Chlosyne harrisii*
Hemlock woolly adelgid	*Adelges tsugae*
Hermit thrush	*Catharus guttatus*
Hickory tussock moth	*Lophocampa caryae*
Hog	*Sus scrofa*
Honeybee	*Apis mellifera*
Indiana bat	*Myotis sodalist*
Ladybug beetles	*Coccinella* spp.
Linden looper	*Erannis tiliaria*
Little brown bat	*Myotis lucifugus*
Llama-like camel	*Camelops* sp.
Long-tailed weasel	*Mustela frenata*
Magnolia warbler	*Setophaga magnolia*
Mammoth	*Mammuthus columbi*
Many-ribbed salamander	*Eurycea multiplicate*
Marbled salamander	*Ambystoma opacum*
Mastodon	*Mammut americanum*
Mountain chorus frog	*Pseudacris brachyphona*
Mountain dusky salamander	*Desmognathus ochrophaeus*
Mountain lion	*Felis concolor*
Mourning dove	*Zenaida macroura*
Musk ox	*Ovibos moschatus*
Nine-banded armadillo	*Dasypus novemcinctus*
Northern cardinal	*Cardinalis cardinalis*
Northern cottonmouth	*Agkistrodon piscivorus*
Northern flying squirrel	*Glaucomyx sabrinus*
Northern long-eared bat	*Myotis septentrionalis*
Northern parula	*Setophaga americana*
Northern spring peeper	*Psudoacris crucifer*
Norway rat	*Rattus norvegicus*
Oklahoma salamander	*Eurycea tynerensis*

Common name	Taxonomic name
Orange-crown warbler	*Leiothlypis celata*
Ozark big-eared bat	*Corynorhinus townsendii* subsp. *ingens*
Ozark hellbender	*Cryptobranchus alleganiensis* subsp. *bishop*
Ozark zigzag salamander	*Plethodon angusticlavius*
Passionflower bee	*Protandrena passiflorae*
Peaks of Otter salamander	*Plethodon hubrichti*
Peregrine falcon	*Falco peregrinus*
Plains spotted skunk	*Spilogale interrupta* (formerly *Spilogale putorius* subsp. *interrupta*)
Prothonotary warbler	*Protonotaria citrea*
Racoon	*Procyon lotor*
Red fox	*Vulpes vulpes*
Red oak borer	*Enaphalodes rufulus*
Red wolf	*Canis rufus*
Red-backed salamander	*Plethodon cinereus*
Red-bellied woodp*ecker*	*Melanerpes carolinus*
Red-cockaded woodpecker	*Picoides borealis*
Red-eyed vireo	*Vireo olivaceous*
Red-headed woodpecker	*Melanerpes erythrocephaus*
Red-tailed hawk	*Buteo jamaicensis*
Redbay ambrosia beetle	*Xyleborus glabratus*
Ridge and valley salamander	*Plethodon hoffmani*
Ringneck snake	*Diadophis punctatus*
River otter	*Lontra canadensis*
Roadrunner	*Geococcyx californianus*
Rock vole	*Microtus chrotorhinus*
Sabre-tooth cat	*Smilodon fatalis*
Short-faced bear	*Arctodus simus*
Southeastern blueberry bee	*Habropoda laboriosa*
Southern Appalachian salamander	*Plethodon teyanalee*
Southern black widow	*Latrodectus variolus*
Southern flying squirrel	*Glaucomys volans*
Southern pine beetle	*Dendroctonus frontalis*
Southern short-tailed shrew	*Blarina carolinensis*
Southern unstriped scorpion	*Vaejovis carolinianus*
Spongy moth	*Lymantria dispar*
Spotted salamander	*Ambystoma maculatum*
Spring beauty miner bee	*Andrena erigeniae*
Striped bark scorpion	*Centruroides vittatus*
Striped skunk	*Mephitis mephitis*
Texas brown tarantula	*Aphonopelma hentzi*)
Teyanalee salamander	*Plethodon teyanalee*
Threetooth snail	*Triodopsis platysayoides*
Tiger wolf spider	*Tigrosa aspersa*

Common name	Taxonomic name
Timber rattlesnake	*Crotalius horridus*
Timber wolf	*Canis lupus*
Tricolored bat	*Perimyotis subflavus*
Tufted titmouse	*Baeolophus bicolor*
Turkey	*Meleagris gallopavo*
Turkey vulture	*Cathartes aura*
Turpentine beetle	*Dendrotonus terebrans*
Veery	*Catharus fuscescens*
Virginia opossum	*Didelphis virginina*
Wehrle's salamander	*Plethodon wehrlei*
Western diamondback rattlesnake	*Crotalus atrox*
Western slimy salamander	*Plethodon albagula*
Western worm snake	*Carphophis vermis*
Whip-poor-will	*Antrostomus vociferus*
White-breasted nuthatch	*Sitta carolinesis*
White-eyed vireo	*Vireo griseus*
White-footed deer mouse	*Peromyscus lecopus*
White-tailed deer	*Odocoileus virginianus*
Wood thrush	*Hylocichia mustelina*
Woodland vole	*Microtus pinetorum*
Worm snake (Appalachians)	*Carphophis amoenus*
Worm snake (Ozarks)	*Carphophis vermis*
Worm-eating warbler	*Helmitherus vermivorum*
Yellow passion flower	*Passiflora lutea*
Yellow warbler	*Setophaga petechia*
Yellow-cheeked vole	*Microtus xanthognathus*
Yellow-spotted woodland salamander	*Plethodon pauleyi*
Fungi and Lichens	
Armillaria root rot	*Armillaria mellea*
Artist's conk polypore	*Ganoderma applanatum*
Beech bark fungus	*Neonectria ditissima*
Beech bark fungus	*Neonectria fagisuga*
Birch polypore	*Piptoporus betulinus*
Black morel	*Morchella angusticeps*
British soldier lichen	*Cladonia cristatella*
Cedar apple gall rust	*Gymnosporangium juniperi-virginianae*
Chestnut blight fungus (current name)	*Cryphonectria parasitica*
Chestnut blight fungus (earlier name)	*Diaporthe parasitica*
Chestnut blight fungus (earlier name)	*Endothia parasitica*
Chinese chestnut	*Castanea mollissima*

Common name	Taxonomic name
Common morel	*Morchella americana*
Common toadstool lichen	*Lasallia mammulata* and *L. pensylvanica*
Deer truffle	*Elaphomyces americanus*
Disease-causing oomycete	*Phytophthora* spp.
Dogwood anthracnose fungus	*Discula destructiva*
Dutch elm disease fungus	*Ophiostoma ulmi*
Dyer's polypore	*Phaeolus schweinitzii*
Earth lichen	*Dermatocarpus lachneum*
Gem-studded puffball	*Lycoperdon perlatum*
Golden moonglow lichen	*Dimelaena oreina*
Hen of the woods	*Grifolia fondosa*
Honey mushroom	*Armillaria mellea*
Laurel wilt disease	*Raffaelea lauricola*
Lion's mane	*Hericium erinaceus*
Magnolia-cone xylaria	*Xylaria magnoliae*
Mayapple rust	*Allodus podophylli*
Oak wilt fungus	*Bretziella fagacearum*
Oak-loving gymnopus	*Gymnopus dryophilus*
Oyster mushroom	*Pleurotus ostreatus*
Pine oak gall rust	*Cronartium quercuum*
Red ring rot fungus	*Phellinus pini*
Red species of Amanita	*Amanita arkansana*
Red-belted conk	*Fomitopsis mounceae*
Reindeer mosses	*Cladina* spp.
Rock tripe lichens	*Umbilicaria* spp.
Root disease fungus	*Heterobasidion* spp.
Saggy mane	*Coprinus comatus*
Salted shield lichen	*Parmelia saxatilis*
Scarlet cup fungus	*Sarcoscypha occidentalis*
Shield lichens	*Parmelia* spp.
Smooth rock tripe	*Umbilicaris mammulata*
Sulphur-shelf polypore	*Laetiporus sulphureus*
Sweet gum fruit fungus	*Xylaria liquidamber*
Tinder fungus	*Fomes fomentarius*
Toadstool lichen	*Melanetia culbertonii*
Tulip morel	*Morchella diminutiva*
Turkey tail fungus (or mushroom)	*Tremetes versicolor*
Velvet-top fungus	*Phaeolus schweinitzii*
White nose syndrome fungus	*Pseudogymnoascus destructans*
White pine blister rust	*Cronartium ribicola*
Yellow morel	*Morchella americana*

Appendix B: General Distribution Patterns of the More Common Trees Found in the Appalachians and Ozarks

		Distribution in the Ozarks	Distribution in the Appalachians
American beech	*Fagus grandifolia*	Common on mesic sites throughout	Uncommon and limited to mesic sites
American chestnut	*Castanea dentata*	Once common throughout but eliminated by the chestnut blight	Not present in the region
Balsam fir	*Abies balsamea*	Restricted to relatively few high-elevation mesic sites	Not present in the region
Basswood	*Tilia americana*	Occasional	Infrequent
Bitternut hickory	*Carya cordiformis*	Common throughout	Common throughout
Black birch	*Betula lenta*	Common throughout	Not present in the region
Black cherry	*Prunus serotina*	Common throughout	Common throughout
Black hickory	*Carya texana*		Occasional
Black locust	*Robinia pseudoacacia*	Common throughout	Common throughout
Black oak	*Quercus velutina*	Common throughout	Common throughout
Black walnut	*Juglans nigra*	Occasional	Occasional
Blackjack oak	*Quercus marilandica*	Occasional	Common throughout
Blue ash	*Fraxinus quadrangulata*	Uncommon	Occasional
Bur oak	*Quercus macrocarpa*	Uncommon	Occasional to common in the northern Ozarks
Canadian hemlock	*Tsuga canadensis*	Occasional but limited to mesic sites	Not present in the region
Chestnut oak	*Quercus montana*	Common throughout	Not present in the region
Chinquapin oak	*Quercus muehlenbergii*	Common throughout	Common
Cucumber magnolia	*Magnolia acuminata*	Common throughout	Infrequent
Fraser fir	*Abies fraseri*	Restricted to relatively few high-elevation mesic sites	Not present in the region
Mockernut hickory	*Carya alba*	Common throughout	Common throughout
Northern red oak	*Quercus rubra*	Common throughout	Common throughout

		Distribution in the Ozarks	Distribution in the Appalachians
Pignut hickory	*Carya glabra*	Occasional	Occasional
Post oak	*Quercus stellata*	Uncommon	Common throughout
Red maple	*Acer rubrum*	Common throughout	Common throughout
Red spruce	*Picea rubens*	Restricted to high-elevation mesic sites	Not present in the region
Scarlet oak	*Quercus coccinea*	Occasional	Common in the northern Ozarks
Shagbark hickory	*Carya ovata*	Common throughout	Common throughout
Shortleaf pine	*Pinus echinata*	Widespread but mostly restricted to xeric sites at lower elevations	Commonly associated with exposed sandstone formations in the Ozarks, widespread and common in the Ouachita Mountains
Southern red oak	*Quercus falcata*	Occasional	Common throughout
Sugar maple	*Acer saccharum*	Common throughout	Common throughout
Sweet gum	*Liquidambar styraciflua*	Occasional	Common throughout
Tulip poplar	*Liriodendron tulipifera*	Common throughout	Uncommon
White ash	*Fraxinus americana*	Common throughout	Common throughout
White oak	*Quercus alba*	Common throughout	Common throughout
Yellow birch	*Betula alleghaniensis*	Restricted to higher elevation moist sites	Not present in the region

Appendix C: Brief Descriptions of the More Common Trees Found in the Appalachians and Ozarks

*American Beech (*Fagus grandifolia *Ehrh.)*

American beech occurs in both the Appalachians and Ozarks. The species is restricted to relatively mesic sites but can be found from the lowest elevations in the two regions to 1500 m in the Southern Blue Ridge. In the Ozarks, the distribution is discontinuous. In Missouri, it is found only in the extreme southeastern part of the state, which is technically not part of the Ozarks, and in Arkansas it is found along the major rivers (e.g., the Buffalo River) and the Ouachita Mountains. American beech has a spreading crown and typically reaches a height of 20–30 m, but exceptional individuals can reach 30 m. It is easily recognized by its smooth and thin gray bark. This species is shade tolerant. The leaves are 12–16 cm long and serrate with nearly parallel pinnately arranged veins. The fruit is a small (10–15 mm), sharply three-angled nut occurring singly or in pairs within a 1.5–2.5 cm long short-spined cupule. In the fall, the leaves turn bronze/brown. Species of beech other than American beech are common in Western Europe, and the name "beech" is an old English word dating back centuries.

American Chestnut (*Castanea dentata* [Marsh.] Borkh.)

Until it was virtually eliminated from the forest canopy by the chestnut blight fungus, American chestnut was one the most abundant and important trees in the Appalachians. However, the tree is still present as sapling-sized individuals in many of the areas (especially those at higher elevations) where the species was once dominant. This tree does not occur in the Ozarks. American chestnut is a large tree, reaching a height of 20–30 m and a diameter of 1–2 m. The leaves are alternate, 14–20 cm long and 7–10 cm wide, oblong-lanceolate, and coarsely serrate, with sharply pointed teeth. The tree produces its fruits (nuts) in a spiny cupule (or bur) 4–10 cm in diameter. The cupule encloses 2–3 nuts, each 1–2.5 cm in diameter and flattened on one side. The nuts are highly edible. American chestnut occurs on all types of sites but the best developed pre-blight forests occurred at mid-slope positions on southern exposures. American chestnut is shade tolerant. In the fall, the leaves turn yellow to yellow-brown. The name chestnut is derived from the early English term "chesten nut."

Balsam Fir (*Abies balsamea* [L.] Mill.)

Balsam fir is an evergreen tree limited to only a few small populations in the Appalachians and is not present in the Ozarks. The populations in the Appalachians are mostly associated with high-elevation wetland sites, including bogs. This is not always the case further north beyond the region considered in this book. Balsam fir is a medium-sized tree in the Appalachians, although in other portions of its range in North America it can reach a height of 15–20 m. The tree has a straight trunk and a conical crown. It is shade tolerant. The bark is thin, gray, and characterized by the presence of "resin blisters." The leaves are flat, 15–30 mm long, and 1.3–1.8 mm wide. The seed cones are resinous, sessile, erect, cylindrical to broadly ovoid, and 40–80 mm long by 15–35 mm wide. The tree derives its name from its fresh fragrance produced by the balsam that is exuded by this tree.

Basswood (*Tilia americana* L.)

Basswood (sometimes called American basswood) is a medium-sized to large tree found in mesic sites throughout the Appalachians and Ozarks but rarely occurring as an important species. The tree can reach a height of 18–35 m and grows more rapidly that most of the species with which it co-occurs. Basswood is shade tolerant, and it is not uncommon for the trunks of two or more basswood trees to be found growing closely together. The leaves are alternate, ovate to cordate, 6–20 cm long and 5–18 cm wide, and have an asymmetrical base. Much larger leaves sometimes occur on young stems. The flowers are small, yellowish-white, 10–14 mm in diameter, and occur in clusters of 6–20 along with a distinctive leaf-like bract. In the fall, the leaves turn yellow to orange. The name basswood is derived from the term "bastwood" and refers to the tough inner bark (or bast) of this tree. Three taxonomic varieties (*americana*, *caroliniana*, and *heterophylla*) have been recognized for *Tilia americana*, but these intergrade and are rarely considered in surveys of forest composition.

Bitternut Hickory (*Carya cordiformis* [Wangenh.] K.Koch)

Bitternut hickory is associated with mesic sites throughout the Appalachians and Ozarks but is rarely more than a minor component of the canopy. This is a medium-sized to large tree that can reach a height of 15–30 m. The leaves are alternate, compound, 15–30 cm long and have 7–9 leaflets. The leaflets are lanceolate, 4–12 cm long, 3–6 cm wide, and the apical leaflet is the largest. The fruit is a subglobose nut, 2–3 cm long, and has a hard, bony husk. The tree derives its name from the bitter-tasting kernel of the nut. The major identifying characteristic of bitternut hickory is the bright sulfur-yellow bud. Bitternut hickory is shade intolerant. In the fall, the leaves turn yellow brown.

Black Birch (*Betula lenta* L.)

Black birch is found throughout the Appalachians but is absent from the Ozarks. This is a medium-sized to large tree that is typically 15–25 m tall. It is found on all types of sites except for the most xeric. Black birch has intermediate shade tolerance. The bark is dark brown to almost black and smooth at first but separating into irregularly shaped plates in older trees. The leaves are alternate, ovate, serrate along the margins, and 6–10 cm long. This tree produces cone-like catkins in which the seeds are produced. The female catkin is persistent, erect, and 2.0–2.5 cm long. This tree derives its name from the color of its bark. In the fall, the leaves turn a bright yellow or gold.

Black Cherry (*Prunus serotina* Ehrh.)

Black cherry is the largest native cherry in North America and is widespread in both the Appalachians and Ozarks. It is commonly associated with moist sites and is probably most abundant in northern hardwood forests in the Allegheny Mountains. This tree can reach a height of 15–25 m and can be recognized by its rough black bark and, in the spring, the loose racemes of small white flowers. Black cherry has intermediate shade tolerance, but few of the numerous seedlings that develop from seeds that germinate beneath the canopy of this tree survive. The leaves are alternate, 5–13 cm long, ovate-lanceolate in shape, and minutely toothed on the margins. They contain a small amount of cyanide and thus are poisonous. The fruit is a round drupe, 8–10 mm in diameter and dark blue or black. The color of the fruit is the basis for the common name of this tree. In the fall, the leaves turn yellow and orange.

Black Locust (*Robinia pseudoacacia* L.)

Black locust is a medium-sized tree that can reach a height of 15–25 m. The species is common throughout the Appalachians and Ozarks but tends to be associated with xeric sites, especially those on which early successional forests occur. The leaves are alternate, compound, 15–35 cm long, and consist of 9–19 rounded leaflets, each of which is 2.5–5 cm long and 1–1.5 cm wide. This tree produces pendant racemes bearing numerous white flowers, each about 2.5 cm wide. The flowers are conspicuous, very fragrant, and produce large amounts of nectar. As a result, they attract large numbers of insects. The fruit is an oblong, flattened legume, 5–10 cm long. Black locust is highly shade intolerant. The tree derives its name from the black or almost black color of the bark. In the fall, the leaves turn yellow. As a member of the Fabaceae, the roots of black locust contain nodules that allow nitrogen fixation to occur.

Black Oak (*Quercus velutina* Lam.)

Black oak is commonly found throughout both the Appalachians and Ozarks. It is a medium-sized tree with an irregular crown and reaching a typical height of 20–30 m. Black oak is usually associated with relatively dry sites. The leaf is 10–20 cm long, obovate or ovate in shape, and has 5–7 bristle-tipped lobes. Leaf shape is highly variable, with leaves directly exposed to the sun (sun leaves) having deep sinuses and leaves in the shade (shade leaves) having very shallow sinuses. Black oak has intermediate shade tolerance but is less tolerant than most of the other trees in the forests in which it occurs. The fruit is an ovoid acorn (1–2.5 cm) one-third to one-half covered by a bowl-shaped cap. The name of this tree is derived its dark blackish or brownish grey bark.

Blackjack Oak (*Quercus marilandica* Uünchh.)

Blackjack oak is characteristic of open xeric sites and occurs at scattered localities throughout the Ozarks but is uncommon in the Appalachians. This is a medium-sized tree that reaches a height of 10–15 m. The most recognizable feature is the bark, which is black, very thick, and rough. The thick bark allows the tree to be highly tolerant of fire. Blackjack oak is shade intolerant. The leaves are alternate, 7–15 cm long and 7–15 cm wide, broadly wedge-shaped, and with three bristle-tipped lobes near the apex. The deep cup encloses one third to one half of the nut, which is ovoid and 10–15 mm long. The name of this tree is derived from the blackjack-like shape of the leaves. In the fall, the leaves turn orange or red.

Blue Ash (*Fraxinus quadrangulata* Michx.)

Blue ash is occasionally found in forests on mesic sites in the northern Ozarks but is rare in the Appalachians. It a medium-sized tree can reach a height of 20–25 m. This tree can be recognized from the 4-angled twigs that are sometimes winged along the ridges. The leaves are opposite, compound, 20–30 cm long, and consist of 5–11 (but mostly 7) leaflets, with each leaflet 7–13 cm long and 2.5–5 cm wide. The leaflets have a serrate margin. This tree has intermediate tolerance to shade. The fruit is a samara, 2.5–5 long. In the fall, the leaves turn a pale yellow. The name of this tree is derived from the blue dye extracted from the inner bark. Like all members of the genus, blue ash is treated by the emerald ash borer.

Bur Oak (*Quercus macrocarpa* Michx.)

Bur oak is found throughout the Ozarks but uncommon to rare in the Appalachians. This species is associated with mesic sites and is a large tree that can reach a height of 15–40 m. Bur oak has intermediate shade tolerance. The leaves are alternate, 7–15 cm long and 3–13 cm wide, obovate or oblong, and there are 3–6 rounded lobes, with the middle sinus the deepest. The deep bowl-shaped cup encompasses one half to three fourths of the nut, which is 1.5–3.5 cm long. The uppermost scales of the cup are elongate and curled, thus resembling a bur. This feature accounts for the common name of this tree. In the fall, the leaves turn yellow to brown.

Canadian Hemlock (*Tsuga canadensis* [L.] Carr.)

Canadian hemlock is a large evergreen tree associated with mesic sites throughout the Appalachians. It is most common along streams and ravines but can occur along with red spruce on upper slopes and ridgetops at higher elevations. Canadian hemlock has conic crown, is typically 20–30 m tall, and can reach a diameter of more than a meter. This is a long-lived tree and individuals >500 years old are known. It is shade tolerant. The leaves are flat, needle-like, 8–13 mm long, have blunt rounded tips, and occur in two ranks. The seed cones are ovoid and 1.5–2.5 mm long. The name hemlock is derived from the Old English word "hemeluc" (meaning tree of grace). The Canadian hemlock is severely threatened throughout its entire range by an insect, the hemlock wooly adelgid.

Chestnut Oak (*Quercus montana* Willd.)

Chestnut oak occurs throughout much of the Appalachians, where it is one of the more ecologically important trees, but is absent from the Ozarks. It is relatively common on xeric sites including ridgetops. Chestnut oak is a medium-sized to large tree that typically reaches a height of 20–30 m. On more mesic sites this tree typically develops a straight trunk and a narrow crown, but on xeric (especially ridgetop) sites the trunk is shorter and often crooked. This species is relatively shade tolerant. The bark is very thick and marked with longitudinal fissures separating more elevated rounded areas. The leaves are alternate, 12–20 cm long and have 10–15 shallow rounded lobes on each side. The fruit is a large ellipsoid acorn (2.5–3.5 cm), with about one-third covered by the cup. In the fall, the leaves turn yellow-brown. This tree derives its name from the resemblance of the leaf-that of American chestnut. Chestnut oak is listed as *Quercus prinus* in older publications.

Chinquapin Oak (*Quercus muehlenbergii*) Engelm

Chinquapin (or chinkapin) oak is found throughout the Appalachians and Ozarks but is much more common in the latter region. It is a medium-sized tree that reaches a maximum height of about 18 m. Chinquapin oak is often associated with drier upper slope positions, rocky glades, and other sites with soils derived from limestone. The leaves are 5–15 cm long and 4–8 cm wide. The fruit is a small oval acorn (1.3–2.5 cm), with about one-fourth or one-half it covered by the cup. Chinquapin oak derives its name from the resemblance of the leaf to that of the chinquapin, which is similar to the leaf of the closely related American chestnut. The leaf of chinquapin oak is similar in shape to the leaf of chestnut oak, but the teeth of the former are pointed and those of the latter are rounded. In the fall, the leaves of chinquapin oak turn yellow, orange, or brown.

Cucumber Magnolia (*Magnolia acuminata* L.)

The cucumber magnolia is widely distributed but rarely a dominant species in the forests of the Appalachians and Ozarks. This tree generally occurs on relatively mesic but well-drained sites and is most abundant in cove forests of the Southern Blue Ridge. It is a large straight-trunked tree that can reach a height of 18–25 m. Cucumber magnolia has intermediate shade tolerance. The leaves are alternate, large (12–25 cm long and 2–12 cm wide) and ovate to widely elliptical in shape. The flowers of cucumber magnolia are large and distinctive, with yellow to yellow-green petals 4–7 cm long. The fruit is a compound follicle, 3–7 cm long and 1–3 cm in diameter. The resemblance of the fruit to a cucumber accounts for the common name of this species. In the fall, the leaves turn a dull yellow.

Fraser Fir (*Abies fraseri* [Pursh] Poir.)

Fraser fir is an evergreen tree characteristic of the very highest elevations in the Southern Blue Ridge, with the northernmost population occurring on Mt. Rogers in southwest Virginia. The species is not present in the Ozarks. Fraser fir is a medium-sized tree but can reach a height of 10–20 m. The tree has a straight trunk and a conical crown. This tree is shade tolerant. The bark is thin, gray, and characterized by the presence of "resin blisters" when the tree is young but the bark becomes scaly with age. The leaves are flat, 10–23 mm long, 1.5–2.0 mm wide, and spirally arranged on twigs. The seed cones are resinous, sessile, erect, cylindrical to broadly ovoid, and 35–70 mm long by 15–38 mm wide. The tree derives its name from the Scottish botanist John Fraser. Like balsam fir, the leaves have a subtle fragrance produced by the exuded balsam. Fraser fir and balsam are very similar, and the former has sometimes been considered as a subspecies of the latter. Fraser fir has been seriously threatened throughout its range by the balsam woolly adelgid.

Mockernut Hickory (*Carya alba* [L.] Nutt. ex Elliott)

Mockernut hickory, also called white hickory, is a medium-sized to large tree commonly found in the forests of both the Appalachians and Ozarks. The tree can reach a height of 20–30 m. The leaves are alternate, compound, 20–50 cm long and consist of 7–9 leaflets, each of which is 5–20 cm long and 2–8 cm wide. The rachis and leaves are densely pubescent, which enables this species to be easily identified from this feature alone. This tree is shade tolerant. The fruit is a pear-shaped, thick-shelled (5–6 mm) nut 3–4.5 long, and containing a small kernel. The hull splits apart at maturity. In the fall, the leaves turn golden-yellow. The common name of this tree appears to be derived from the struggle one would have to crack the thick husk only to find a small kernel inside. This species is listed as *Carya tomentosa* in some taxonomic treatments.

Northern Red Oak (*Quercus rubra* L.)

Northern red oak is one of the more widespread and ecologically important trees in the Appalachians and Ozarks. It is relatively more common on mesic sites but can occur on virtually any type of site. The species sometimes forms nearly monodominant communities at upper slope positions in some parts of the Appalachians. Northern red oak is a medium-sized tree with a broad spreading crown and a typical height of 20–30 m. It has intermediate shade tolerance but rarely becomes established under a complete canopy. The leaves are alternate, 12–20 cm long and have 7–11 sharply pointed tips. The fruit is a large ovoid to ellipsoid nut (2–3 cm) with a very shallow cup. In the fall, the leaves turn brick-red. The name of the tree is derived from this feature.

Pignut Hickory (*Carya glabra* [Mill.] Sweet)

Pignut hickory is found throughout the Appalachians and Ozarks but is never especially abundant. It can occur on most types of sites. It is a large tree and can reach a height of 20–30 m. The leaves are, alternate, compound, 15–25 cm long, and consist of 5–7 leaflets. Each leaflet is 4–20 cm long and 2–8 cm wide. This tree has intermediate tolerance to shade. The nut is pear-shaped, 2.5–3.5 cm long, and the husk is thick and not ridged. In the fall, the leaves turn golden to rusty yellow. The common name is derived from the fact that early settlers allowed pigs to roam free and eat hickory nuts, often those of this species.

Post Oak (*Quercus stellata* Wangenh.)

Post oak is found on xeric sites throughout the Ozarks and Appalachians but is more common in the former region. It is a small tree, usually no more than 10–15 m tall. Post oak is intolerant of shade. The branching pattern of this tree often gives it a rugged appearance. The leaves are alternate, leathery, densely pubescent beneath, and have three perpendicular terminal lobes that cause it to resemble a cross. The nuts are 1.5–2.5 cm long and one half to one fourth enclosed by the cup. In the fall, the leaves turn golden brown. The name of this tree is derived from its frequent use to make fence posts.

Red Maple (*Acer rubrum* L.)

Red maple occurs throughout both the Appalachians and Ozarks but is rarely a dominant species. It can be found on all types of sites. This tree typically reaches a height of 12–20 m but can grow to 30 m. However, most of the time in both regions it is smaller. Red maple is shade tolerant. The leaves are opposite, 8–15 mm long and about the same width. There are three (more rarely five) lobes and the lower surface of the leaf is lighter in color than the upper surface. The fruit is a pair of samaras 1.8–2.4 long and with conspicuous wings. In the fall, the leaves turn a bright crimson. The latter accounts for the name of this tree.

Red Spruce (*Picea rubens* Sarg.)

Red spruce is a large evergreen tree that is usually associated with high elevation upper slopes and ridgetops in the Appalachians but is absent from the Ozarks. It also can occur in some wetland habitats such as bogs. This tree can reach a height of 20–25 m and a diameter of almost a meter but is much smaller on exposed, ridgetop sites. Red spruce is very shade tolerant and can form nearly impenetrable thickets of tightly packed small trees after logging. The leaves are 12–15 mm long, 4-sided, slightly curved, and have a sharp point. The seed cones are pendant, cylindrical, and

3–5 cm long. Red spruce derives its name from its reddish-brown bark, with the reddish color more apparent beneath the scales that form the outer layer of the bark. Red spruce is long-lived, with individual trees known to exceed 400 years.

Scarlet Oak (*Quercus coccinea* Muench.)

Scarlet oak occurs on xeric sites throughout the Appalachians but is restricted largely to the northern portion of the Ozarks. It is a large tree that can reach a height of 20–30 m. Scarlet oak is shade intolerant and is most likely to be found in an early successional forest. The leaves are alternate, ovate or obovate, have 5–9 lobes with wide sinuses that are nearly round, and are 7–16 cm long and 8–13 cm wide. The cup is bowl-shaped, enclosing one third to one half of the nut, the latter 1.5–2.2 cm long and usually with several concentric circles at the apex. In the fall, the leaves turn a brilliant scarlet. The name of the tree is derived from this feature.

Shagbark Hickory (*Carya ovata* [Mill.] K.Koch)

Shagbark hickory is found throughout the Appalachians and Ozarks but is not especially common. This is a large tree, reaching a height of 20–40 m. The leaves are alternate, compound, and 20–35 cm long, with 5 (sometimes 7) leaflets, the latter 8–22 cm wide and 2.5–9 cm wide. The bark of shagbark hickory exfoliates in long plates, which is the tree's most distinctive feature. Shagbark hickory has intermediate tolerance to shade. The nut is globose 3–6 cm long, and sometimes slightly flattened on one side. In the fall, the leaves turn yellow to orange. Shagbark hickory derives its name from the exfoliating plates of bark, which give the tree a "shaggy" appearance.

Shortleaf Pine (*Pinus echinata* Mill.)

Shortleaf pine is a large evergreen tree characteristic of xeric, often sandy or rocky sites throughout the Appalachians and Ozarks. It is the only pine associated with the Ozarks of Missouri, where it can be found in monodominant forests as well as in woodlands along with oaks. Shortleaf pine is typically 15–30 m tall, with a straight trunk usually with numerous secondary branches. This tree is shade intolerant and requires bare soil in open sunlight to reproduce. The leaves are needle-like, 7–13 cm long, and occur mostly in 2's but sometimes in 3's. In older trees, the bark tends to break into more or less rectangular plates. The seed cone is ovoid to conic and about 5 cm long. The name is based on the relative length of the needles compared to other species of pine that occur in the same region of the United States.

Southern Red Oak (*Quercus falcata* Michx.)

Southern red oak is associated with xeric sites in the southern Ozarks but has a limited distribution in the Appalachians. It is a large tree that typically reaches a height of 20–30 m. Southern red oak has intermediate shade tolerance. The leaves are alternate, obovate to ovate, have 3–5 lobes, with the lowermost lobes much larger than the others. The leaves are 10–22 cm long and 6.5–16 cm wide. The nut is 10–12 mm long and about one-third enclosed by the cup. In the fall, the leaves turn red to reddish brown. The fall foliage and its general distribution in the United States account for its common name.

Sugar Maple (*Acer saccharum* Marsh.)

Sugar maple is characteristic of mesic sites throughout the Appalachians and the Ozarks. It is a large tree with a usual maximum height of 30–35 m. Sugar maple is shade tolerant. The leaves are opposite, 8–15 cm long and about the same width, 3–7 lobed, and have rounded sinuses. The fruit consists of paired samaras 2.3–4.0 cm long and with conspicuous broad wings. In the fall, the leaves turn a brilliant yellow or scarlet. The tree has long been known as a source of maple syrup and maple sugar, which accounts for its name. However, its use as a source of sugar is restricted largely to some areas of the Appalachians.

Sweet Gum (*Liquidambar styraciflua* L.)

Sweet gum is found on relatively mesic sites throughout the southern Ozarks but has a limited distribution in the northern Ozarks and Appalachians. This is a medium-sized to large tree, typically reaching 15–20 m but is occasionally up to 40 m tall. Sweet gum is intolerant of shade. The leaves are alternate, 8–13 cm wide and have five (occasionally 3 or 7) sharply pointed lobes, thus appearing star-shaped. The fruit is hard, dry, and globose structure 2.5–4.0 cm in diameter. In the fall, the leaves turn to colors ranging from yellow to red to purple. The name of the tree refers to the sweet-tasting gum it produces.

Tulip Poplar (*Liriodendron tulipifera* L.)

Tulip poplar is a large (and sometime very large) tree found throughout the Appalachians but which has a very restricted distribution in the Ozarks. It is fast-growing tree that is typically 10–25 m tall but exceptional individuals can reach 50 m and have a diameter of 1.5–2.5 m. Tulip poplar is associated with moist sites and can be the overwhelming dominant in early successional forests. This tree is essentially shade intolerance but can survive in partial shade. The leaves are alternate, truncate or broadly notched, and 12.5–15.0 cm both long and wide. They have four lobes and are slightly wedge-shaped at base. Tulip poplar has large, conspicuous flowers with petals 5 cm long and greenish yellow in color except at the base, which is orange. The fruit is a narrow light brown cone-like structure formed from numerous overlapping samaras that are dispersed by wind, leaving behind a persistent axis. In the fall, the leaves turn a golden-yellow. Yellow poplar derives its name from the resemblance of its flowers to those of tulips.

White Ash (*Fraxinus americana* L.)

White ash occurs on mesic sites throughout the Appalachians and Ozarks. It is medium-sized to large tree, reaching a height of 20–40 m. The dark gray bark with interlacing deep fissures is distinctive. White ash has intermediate shade tolerance. The leaves are opposite, compound, 20–30 cm long, and consist of 5–9 (but usually 7) leaflets, each 6–13 cm long and 3–6 cm wide. The fruit is a flattened samara 2.5–3.5 long. In the fall, the leaves turn yellow, red, or sometimes purple. The name white ash derives from the dull grayish-green color of the lower surface of the leaves. Like other species of ash, white ash is threatened by the emerald ash borer.

White Oak (*Quercus alba* L.)

White oak is one of the more widespread and common trees in the Appalachians and Ozarks but is most abundant throughout the Ozarks. It occurs on all types of sites but like most trees grows best on deep, moist, well-drained soils. White oak is most competitive on drier sites where it sometimes co-occurs with pine. White oak is a large tree with a wide round to spreading canopy. This species typically reaches a height of 20–30 m and can often have a diameter of more than a meter. White oak has intermediate shade tolerance. The leaves are 12–10 cm long and have 7–9 moderately to deeply divided rounded, finger-like lobes and a wedge-shaped base. The fruit is an oval acorn (2–3 cm), with about one fourth it covered by the cap. In the fall, the leaves turn reddish-brown. White oak derives its name from the ash-gray color of the bark. This tree is long-lived can reach an age of more than 400 years.

Yellow Birch (*Betula alleghaniensis* Britt.)

Yellow birch is associated with moist high-elevation, often rocky, sites in the Appalachian, but the species is absent from the Ozarks. This is a large tree that is typically 20–30 m tall. Yellow birch has intermediate shade tolerance. The tree derives its name from its yellowish bark, although the bark also can be silvery gray. A distinctive feature is that the bark tends to separate into horizontal linear strips that curl up on the ends. The leaves are alternate, ovate, serrate along the margin, and 6–10 cm long. This tree produces catkins in which the seeds are produced. The female catkin is persistent, erect, and 2.0–2.5 cm long. This tree derives its name from the color of its bark. In the fall, the leaves turn bright yellow.